T0177345

On Shifting Foundations

RGS-IBG Book Series

For further information about the series and a full list of published and forthcoming titles please visit www.rgsbookseries.com

Published

On Shifting Foundations

State Rescaling, Policy Experimentation and Economic Restructuring in Post-1949 China

Kean Fan Lim

WILEY

This edition first published 2019

The right of Kean Fan Lim to be identified as the author of this work has been asserted in accordance with law.

Registered Office(s)
John Wiley & Sons, Inc., 111 River Street, Hoboken, NJ 07030, USA
John Wiley & Sons Ltd, The Atrium, Southern Gate, Chichester, West Sussex, PO19 8SQ, UK

Editorial Office
9600 Garsington Road, Oxford, OX4 2DQ, UK

For details of our global editorial offices, customer services, and more information about Wiley products visit us at www.wiley.com.

Wiley also publishes its books in a variety of electronic formats and by print-on-demand. Some content that appears in standard print versions of this book may not be available in other formats.

Library of Congress Cataloging-in-Publication Data

Names: Lim, Kean Fan, author.
Title: On shifting foundations : state rescaling, policy experimentation and
   economic restructuring in post-1949 China / Kean Fan Lim.
Description: Hoboken, NJ : John Wiley & Sons, Inc, 2019. | Series: RGS-IBG book series |
   Includes bibliographical references and index. |
Identifiers: LCCN 2018045469 (print) | LCCN 2018053740 (ebook) |
   ISBN 9781119344575 (Adobe PDF) | ISBN 9781119344582 (ePub) |
   ISBN 9781119344551 (hardcover) | ISBN 9781119344568 (pbk.)
Subjects: LCSH: Regional economics–China. | Rural development–China. |
   China–Economic policy–1949– | China–Politics and government–1949–
Classification: LCC HT395.C55 (ebook) | LCC HT395.C55 L56 2019 (print) |
   DDC 338.951–dc23
LC record available at https://lccn.loc.gov/2018045469

Cover Image: Entrance to Hengqin New Area, 2013 © Kean Fan Lim
Cover Design: Wiley

Set in 10/12pt Plantin by SPi Global, Pondicherry, India

The information, practices and views in this book are those of the author(s) and do not necessarily reflect the opinion of the Royal Geographical Society (with IBG).

Printed in Singapore by C.O.S. Printers Pte Ltd

10   9   8   7   6   5   4   3   2   1

For Stephanie and Ethan Lim

Contents

Series Editor's Preface

The RGS-IBG Book Series only publishes work of the highest international standing. Its emphasis is on distinctive new developments in human and physical geography, although it is also open to contributions from cognate disciplines whose interests overlap with those of geographers. The Series places strong emphasis on theoretically-informed and empirically-strong texts. Reflecting the vibrant and diverse theoretical and empirical agendas that characterize the contemporary discipline, contributions are expected to inform, challenge and stimulate the reader. Overall, the RGS-IBG Book Series seeks to promote scholarly publications that leave an intellectual mark and change the way readers think about particular issues, methods or theories.

For details on how to submit a proposal please visit:
www.rgsbookseries.com

David Featherstone
University of Glasgow, UK

RGS-IBG Book Series Editor

Acknowledgements

This book is the culmination of a decade-long research journey. It started with the research proposal I submitted for my graduate school application to the University of British Columbia (UBC) in 2008 – a proposal that enabled me to work with two of the best economic geographers in the world, Trevor Barnes and Jamie Peck, but one that was also *very different* from this book's focus on state rescaling in China. Indeed, it was during the summer of 2009, as I began reading several books by David Harvey that Trevor lent me, that I became interested in China's growing influence in the global system of capitalism. Even so, I did not consider taking on the challenging task of examining the fast-changing economic geographies in China until it was time to draft my dissertation research proposal in the spring of 2011. An emerging pattern of economic–geographical transformations in China piqued my interest at the time: the designation of 'nationally strategic new areas' as frontiers of policy experimentation. Both Trevor and Jamie were particularly encouraging when I mentioned I would like to shift my research focus to these territories – their subsequent patience and support enabled me to develop a feasible proposal that ultimately led to a highly-rewarding research project.

The Editor of the RGS-IBG Book Series, David Featherstone, provided encouraging support throughout the process of writing this book. Thank you for believing in this project and for providing solid editorial guidance that steered the book to its completion. Special thanks also goes to the two academic reviewers for their detailed readings of earlier drafts – the final manuscript has benefited immensely from their comments and suggestions. Indeed, the revision process was enjoyable because it felt like an engaging conversation. Ensuring everything worked seamlessly the moment work started on this project was Jacqueline Scott at Wiley-Blackwell. Her attention to detail, incredible efficiency and kind patience is gratefully acknowledged.

I am indebted to an indispensable group of people who formed the spine of this project – my hosts and respondents in Beijing, Chongqing, and the Pearl

River Delta. Professor Weidong Liu from the Chinese Academy of Sciences (CAS) generously provided an academic home during my fieldwork in Beijing and offered tremendous support during my fieldwork in Chongqing. Even though we spent a limited amount of time together in Chongqing, I learnt a lot from Guoqing Yin's deep knowledge of southwestern China. My respondents were very generous with their time and were very willing to assist with gathering more data. Specific mention goes to an academic respondent (and now friend) in Shenzhen who pointed me to many useful sources published in Mandarin. My former schoolmate from the National University of Singapore and now esteemed professor in geographical political economy at the University of Oregon, Xiaobo Su, was ever ready to connect me to contacts in China when I was setting up my fieldwork. Thank you so much to everyone for helping me.

The Economic Worlds research cluster at the School of Geography in the University of Nottingham, which I joined in the summer of 2014 after completing my doctoral programme, provided a warm environment to develop the book proposal. Special thanks goes to Andrew Leyshon, Louise Crewe, Sarah Hall and Shaun French for all the support and encouragement during my four years in the School. Follow up fieldwork in 2015 was enabled by the School's Research Capacity and Networking Fund, while Elaine Watts provided excellent cartographic assistance. My former academic supervisor at the National University of Singapore, Henry Yeung, very kindly shared his experience of book publishing as I was working on my proposal. Jim Glassman, my doctoral dissertation committee member at UBC, provided valuable advice at every stage of my doctoral training, and the book continues to benefit from his extensive knowledge of East Asian development. The final stage of the writing process took place after I joined the Centre for Urban and Regional Development Studies (CURDS) in Newcastle University in the spring of 2018, and I am very thankful for the support to work on revising and ultimately submitting the manuscript.

Some materials from the book were significantly reconstituted from four single-authored publications in international peer-reviewed academic journals. Chapter 2 was re-written on the basis of the geographical–historical review presented in 'On the shifting spatial logics of socioeconomic regulation in post-1949 China', *Territory, Politics, Governance* 5, no. 1 (2017), pp. 65–91. The framework developed in Chapter 3 was originally published as an agenda-setting paper in the Urban and Regional Horizons section of *Regional Studies*, 51, no. 10, pp. 1580–1593. This paper is entitled 'State rescaling, policy experimentation and path dependency in post-Mao China: a dynamic analytical framework'. Empirical data on the Pearl River Delta in Chapter 4 was published earlier in ' "Emptying the cage, changing the birds": state rescaling, path-dependency and the politics of economic restructuring in post-crisis Guangdong' *New Political Economy* 21, no. 4 (2016), pp. 414–435. Chapter 7's analysis of Chongqing's socioeconomic reforms draws from empirical data published in 'Spatial egalitarianism as a social "counter-movement": On socio-economic reforms in Chongqing.' *Economy and*

Society 43, no. 3 (2014), pp. 455–493. All four articles were published by Taylor and Francis and have been reproduced in this book with kind permission.

Last but not least, I am eternally grateful to my wife, Stephanie Lim, for her unrelenting support over this decade. It is not easy to put everything behind in Singapore and embark on a new journey in two continents, but she has done it in the optimistic belief that I will do well. She has, in all sense of the word, always been there for me – cooking meals with love, going on long walks, and, most crucially, telling me not to give up in the face of challenges. Shortly after work on this book began in the summer of 2016, our lovely son, Ethan Lim, was born in Nottingham. Both Stephanie and Ethan were my sources of motivation and strength in the process of completing the manuscript. This book is dedicated to them.

Chapter One
Introduction

During a keynote address to global leaders at the 2016 B20 meeting in Hangzhou, the Chinese President, Xi Jinping, emphasised how China's developmental approach remains predicated on 'crossing the river by feeling for stones' (*mozhe shitou guohe* 摸着石头过河).[1] While the application of this metaphor is not novel, its recurring reference by the Communist Party of China (CPC) more than 60 years after its introduction by Chen Yun, the Vice Premier of the first governing regime led by Mao Zedong, is noteworthy.[2] Chen advocated a measured approach to change during the early 1950s after China entered an entirely new historical phase as a nation-state – hence the term 'new China' – and could not rely on past experiences for guidance. All the newly-victorious CPC knew was what it did *not* want, namely the inherited institutions associated with feudalism, imperialism and bureaucratic capitalism. When Mao's economic programs failed to meet expectations after three decades of 'transition to socialism' (*shehui zhuyi guodu* 社会主义过渡), Chen (1995: 245; author's translation) insisted again on a tentative approach to change in December 1980: 'We want reforms, but also firm steps…which means "crossing the river by feeling for stones." The steps should be small initially, the movement gradual'. Deng Xiaoping, then newly appointed as paramount leader of the CPC, fully endorsed Chen's exhortation as 'our subsequent guiding agenda' (Deng 1994a: 354; author's translation). Read against this 'agenda', Xi's reference to 'feeling for stones' in Hangzhou almost four decades later raises a theoretically-significant question on post-1949 Chinese

On Shifting Foundations: State Rescaling, Policy Experimentation and Economic Restructuring in Post-1949 China,
First Edition. Kean Fan Lim.
© 2019 Royal Geographical Society (with the Institute of British Geographers).
Published 2019 by John Wiley & Sons Ltd.

political–economic evolution: if the *necessity* to feel for 'stones' indicates a preference for stable foundations within the capricious 'river' of global economic integration, (how) have these foundations shifted from their Maoist origins?

Some answers, this book argues, could be derived from an emergent geographical trend of 'feeling for stones' across China – the intensifying institution of experimental socioeconomic policies within territories designated as 'nationally strategic new areas' (*guojia zhanlüe xinqu* 国家战略新区). Newly established regulatory authorities in these intra-urban territories have been delegated the power to 'move first and experiment first' (*xianxing xianshi quan* 先行先试权) with exploratory reforms deemed to be of national significance. These reforms have become integral to the legitimacy of the CPC – and in particular its socialistic rule – as it negotiates the demands of global economic integration (Lim 2014). To be sure, the demarcation of urban frontiers to drive national-level reforms is not a policy innovation per se; it could be argued that the first wave of marketising reforms in four 'Special Economic Zones' (SEZs) – Shantou, Shenzhen, Xiamen and Zhuhai – generated more transformative impacts on China's developmental pathway during the post-Mao era. After all, the SEZs set in motion the collective willingness to welcome foreign capital, relinquish the Maoist notion of self-sufficiency and tap into what was otherwise idling rural surplus labour. This said, there was very little between the SEZs by way of policy differentiation or positioning within the global economy.[3] On the contrary, a distinguishing feature of this recent series of 'nationally strategic' experimentation is the considerable expansion of its territorial platforms, policy scope, and socioeconomic spheres of influence.

First designated was Pudong New Area in Shanghai. Approved in 1990, the territory has since transformed into a world-renowned city-regional 'motor' – or 'dragon head' (*longtou* 龙头), in popular parlance – of China's economic growth.[4] Subsequent experimentation only (re)gained intensity in 2006, however, after the Hu Jintao regime assigned 'nationally strategic' status to the Binhai industrial region in Tianjin. Three more similar territories were instituted during Hu's tenure, namely the Liangjiang New Area in Chongqing; the Nansha New Area in Guangzhou, which has since been co-opted into a broader Guangdong Free Trade Zone (GFTZ) that includes two other zones previously also termed 'nationally strategic', Hengqin and Qianhai; and the Zhoushan Archipelago New Area off the coast of Zhejiang province. The pace and geographical spread of experimentation grew after Xi Jinping took over the CPC leadership in 2013. At the time of writing, the Xi regime officially assigned 'nationally strategic' status to 13 additional 'new areas' across all major regions in China (see Figure 1.1). These are, namely, Guian New Area, Xixian New Area, Qingdao Xihaian New Area, Dalian Jinpu New Area, Chengdu Tianfu New Area, Changsha Xiangjiang New Area, Nanjing Jiangbei New Area, Fuzhou New Area, Yunnan Dianzhong New Area, Harbin New Area, Changchun New Area, Nanchang Ganjiang New Area and Baoding Xiongan New Area. Viewed holistically, this geographical trend suggests the desire to seek out new 'stones' have never been stronger than at any other

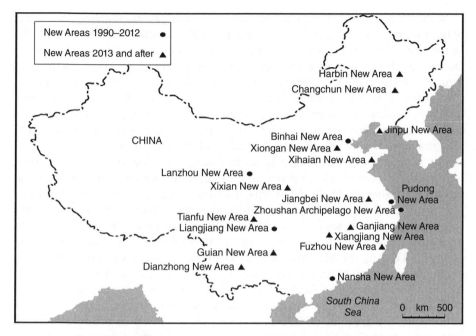

New Areas 1990–2012 ●
New Areas 2013 and after ▲

Harbin New Area ▲
Changchun New Area ▲
Jinpu New Area ▲
CHINA
Binhai New Area ●
Xiongan New Area ▲
Xihaian New Area ▲
Lanzhou New Area ●
Xixian New Area ▲
Pudong New Area ●
Jiangbei New Area ▲
Zhoushan Archipelago New Area ●
Tianfu New Area ▲
Liangjiang New Area ●
Ganjiang New Area ▲
Xiangjiang New Area ▲
Fuzhou New Area ▲
Guian New Area ▲
Dianzhong New Area ▲
Nansha New Area ●
South China Sea
0 km 500

Figure 1.1 New frontiers of reforms: China's 'nationally strategic new areas', 1990–2016. Source: Author, with cartographic assistance by Elaine Watts.

stage of 'crossing the river'. *On Shifting Foundations* aims to explain and evaluate this phenomenon.

At one level, this division and differentiation of Chinese state spatiality could be construed as a proactive attempt on the part of the CPC to engage with the global system of capitalism through its own variant of instituted uneven development. Simultaneously, however, the growing pace of change is *symptomatic* of increasingly severe strains within the national regulatory structure. During the build-up to China's 12th Five-Year Plan (2011–2015), the Chinese central government issued an unprecedented admission that its GDP-focused developmental approach of the past three decades was undertaken in tandem with 10 structural challenges (delineated in Table 1.1). One prominent example can be seen in the extensive extraction of natural resources and low-cost dumping of waste into the biosphere in the pursuit of GPD growth. Similarly, the rollback in rural welfare provision and municipal governments' corresponding denial of (already-minimum) social benefits to rural residents who migrate into and support urban economies generated huge savings that were consequently re-directed into capital-friendly, supply-side projects (Oi 1999; Whiting 2001; He and Wu 2009). Accompanying this rollback was the proliferation of social contradictions that collectively exemplify the fragile social contract that constituted the so-called Chinese economic growth 'miracle'.

Rather than tackle these challenges head on at the national scale, central policymakers chose to develop and test potential solutions within each of the designated 'new areas'. Herein lies a key relationship that will be further examined in this book: the institution of 'nationally strategic' experimental policies through territorial reconfigurations. Specifically, the built environment, administrative boundaries and industrial compositions of targeted city-regions have been repurposed to generate new conditions for reforms. Each 'new area' is charged with experimenting with a predetermined range of national-level initiatives that have been formulated with local conditions in mind.

For instance, experimental policies in Liangjiang New Area in Chongqing built on the broader regional program to develop the western interior (more on this program shortly). To facilitate this, the Chongqing government deepened its reform of another national-level institution – the urban–rural dual structure (*chengxiang eryuan jiegou* 城乡二元结构). This was and remains a direct attempt to dismantle a longstanding and highly discriminatory national institution established during the Mao era – the *hukou*, or household registration, system of demographic controls. In the Pearl River Delta (PRD) extended metropolitan region where Hengqin, Qianhai and Nansha New Areas are located, reforms were focused primarily on financial innovation (*jinrong chuangxin* 金融创新), particularly the creation of 'backflow' channels for RMB in offshore financial centres to 'return' to China (*renminbi huiliu* 人民币回流). These reforms deal with another problematic institution of the Mao era, namely the fixed currency exchange rate system. As trade with the global economy expanded, Chinese merchants developed an over-reliance on the US dollar for trade settlements (Lim 2010; McKinnon 2013). The CPC consequently began promoting the external circulation of RMB through currency swaps, bond issues and trade settlements to reduce dollar usage. Opportunities must be offered to foreigners to

Table 1.1   China's 10 socioeconomic challenges, identified in the proposal of the 12th Five-Year Plan.

1. Increasing constraints of resource environments
2. Relationship between investment and consumption is unbalanced
3. Income distribution gap widened
4. Scientific and technical innovation capacity remains weak
5. Asset structure is unsatisfactory
6. Thin and weak agricultural foundation
7. Lack of coordination in urban–rural development
8. Coexistence of contradictory economic structure and employment pressures
9. Apparent increase in social contradictions
10. Persistent structural and systemic obstacles to scientific development

Source: Suggestions on the 12th Five-Year Plan by the Communist Party of China (p. 3, Mandarin document; NDRC 2011). Author's compilation and translation from Mandarin.

re-invest their RMB holdings in order to expand offshore RMB demand (hence the focus on providing 'backflow channels'). The unfolding of these geographically differentiated sets of new policies raises a question of policy experimentation: to what extent does it lead to foundational institutional change at the national level?

To address this question, this book draws from and advances the current body of research that underscores the importance of policy experimentation under CPC rule. As Sebastian Heilmann's work (2008: 2) shows, 'existing, and initially deficient, institutions can be put to work, transformed, or replaced for economic and social development in an open-ended process of institutional innovation that is based on locally generated solutions rather than on imported policy recipes'. This process is not as top–down and rigid as it appears, observe Heilmann and Elizabeth Perry (2011). On the contrary, the preferred modus operandi is to effect 'adaptive governance' through a type of 'guerrilla-styled' decision-making that originated from the CPC's sporadic and opportunistic military strategy in its 'long revolution' through the 1930 and 1940s. The 'wartime base areas' formula of encouraging decentralised initiative within the framework of centralised political authority proved highly effective when redirected to the economic modernisation objectives of Mao's successors' (Heilmann and Perry 2011: 7). To this it could be added that the 'guerilla' approach was never jettisoned during the Mao era. Indeed, despite Mao's grandiose efforts to institute a Soviet-styled central planning system, socioeconomic regulation was more accurately characterised by decentralised rule in the rural 'People's Communes' (*renmin gongshe* 人民公社), within which more than 80% of the population resided. It was also during this era that senior CPC cadres such as Peng Dehuai, Liu Shaoqi and Li Fuchun called repeatedly for gradualism and pragmatism (including the selective retention of market-based practices that Mao found unacceptable), two defining characteristics of the party's impromptu wartime forays.[5]

Confronting growing external debt and persistent domestic poverty at the time of Mao's passing in 1976, the CPC knew it had to make changes to preclude social and political chaos. It was uncertain, however, on what directions to take without undermining its Marxist–Leninist foundations and its ideological commitment to facilitate the previously-mentioned transition to socialism.[6] This uncertainty re-accentuated the need to 'feel for stones'. As Deng Xiaoping explained in 1978, the end-goal of reform would be the modification of national-level institutions through place-specific experimentation. No pre-existing playbook guided this potentially uncomfortable process:

> Before a unified national agenda is developed, new methods can be launched from smaller parts, from one locality, from one occupation, before gradually expanding them. The central government must allow and encourage these experiments. All sorts of contradictions will emerge during experimentation, we must discover and overcome these contradictions in time. (Deng 1994b: 150; author's translation)

Implemented across a growing number of institutionally-distinct territories, Deng's experimental approach in the 1980s and early 1990s was *qualitatively* distinct from those instituted during the Mao era. While it was superficially like decentralised governance in the People's Communes, experimentation after 1978 strongly encouraged spontaneity while it delicately accommodated Mao-era practices. Thomas Rawski (1995: 1152) puts this shift in clear perspective:

> China's reforms typically involve what might be termed 'enabling measures' rather than compulsory changes. Instead of eliminating price controls, reform gradually raised the share of sales transacted at market prices. Instead of privatization, there was a growing range of firms issuing shares. Production planning does not vanish, but its span of control gradually shrinks. This open-ended approach invites decentralized reactions that the Centre can neither anticipate nor control.

Viewed in relation to the most recent wave of policy experimentation in 'nationally strategic new areas', this 'open-ended' approach continues to define the contemporary spatial logics of socioeconomic regulation in China. 'Crucially', Jamie Peck and Jun Zhang (2013: 380) argue, this approach 'has meant that endogenous state capacities and centralised party control have been maintained through China's developmental transformation'. Then again, reforms in China remain, in Zhichang Zhu's (2007) observation, 'without a theory'. There is, specifically, no explanation why experimental reforms have not led to federal-styled autonomy for subnational governments. Likewise, it is not clear whether increasingly differentiated subnational initiatives could drive national-level institutional change in a way that enhances stability. *On Shifting Foundations* will address these gaps in the following three ways.

First, the book problematises dichotomous portrayals of post-Mao economic development as outcomes of decentralisation and its corollary, uneven development, while the Mao era was characterised by a highly centralised political economy that was committed to socio-spatial egalitarianism. Presenting a fresh conceptual and historical appraisal of the spatial logics of socioeconomic regulation since the founding of contemporary China in 1949, this book argues that the apparent 'downscaling' of governance to city-regional levels since the 1980s has not been a linear, one-track process. What is emerging is at once a further fragmentation of regulatory territories as well as a repurposing of Mao-era institutional foundations. In other words, the on-going round of 'nationally strategic' policy experimentation exemplifies the limits *and* legacies of past socio-spatial configurations, which makes it necessary to historicise the rationale for experimentation in each 'new area'.

Second, the book goes beyond assuming the territorial demarcations of 'nationally strategic' experimental sites as straightforward anointments from the central government.[7] Emerging evidence suggests, indeed, that the establishment of these new zones are outcomes of competitive lobbying by subnational actors. What counts as a 'nationally strategic' policy is therefore a fluid and actively contested entity; the designation of a site for national-level reforms is politicised rather than preordained. Lu Dadao, a senior economic geographer at the Chinese Academy of Sciences and a longstanding consultant to CPC policymakers, candidly describes how this lobbying has complicated the notion and value of the 'national strategy':

> What is the national strategy? It refers to the guiding capacity of the development of one region within the broad national domain, including a huge supportive impact. Currently, at the demands of different regions, the problem is the designation of regions as nationally strategic when they should not be. [...] The current problem is some regions emulate one another to develop plans. Through communication and other manoeuvres, they hope the National Development and Reform Commission [hereafter NDRC, also known as *fagaiwei* 发改委] will organise and draw up these plans and forward them for approval by the State Council. Some of these have been approved, others are still lobbying the central government to organise and include their plans within the national strategy. From the national vantage point, how can it work if everyone is a strategic hub? (Lu, interview with *Liaowang*, 17 June 2010; author's translation)

Xiao Jincheng, the president of the NDRC's Institute of Regional Economy, explains the rationale of this competitive alignment with the 'national strategy':

> There is a particular point of view that so long as the State Council approves a plan for a region, this region will be an important developmental node that will enjoy abundant support through state policies and capital. Hence it is possible to approve many projects and there will be no limits to future development. (Xiao, interview with *Liaowang*, 17 June 2010; author's translation)

These observations by Lu and Xiao collectively suggest that policy experimentation has evolved into a multi-scalar *political* process. This evolution exemplifies the *willingness* within the central leadership to consider proposals from subnational actors 'through communication and other manoeuvres'. Integral to the reconfiguration is the alignment of supposedly favourable local conditions with national-level concerns. Why and how some subnational cadres succeed in convincing central policymakers to 'scale up' their territories as 'nationally strategic' will be further explored in this book's two case studies on the PRD and Chongqing (see, specifically, Chapters 4 and 6).

Last, but not least, this book will evaluate the national impact of the experimental policies in the PRD and Chongqing (Chapters 5 and 7, respectively). At least five years have passed since policy experimentation was instituted in these territories, and the CPC Party Secretaries who were directly involved in their designation – Wang Yang and Bo Xilai – have left their positions. It would be apt, then, to evaluate whether the policies have generated fresh avenues for changes in national-level institutional foundations. The focus will be on two major policies in the case study of Chongqing – urban–rural integration and state-driven attempts to attract capital inland from the coastal provinces. In the PRD, the book evaluates the effectiveness of RMB backflow channels in Hengqin and Qianhai New Areas, and foregrounds their implications for Nansha New Area (which was still new at the time of data collection and hence difficult to ascertain policy effectiveness). In so doing, these chapters address the concerns of Lu and Xiao by showing the extent to which 'nationally strategic' designation can lead to foundational change at the national level.

To further contextualise the three foregoing research avenues, the next section will consider how recurring spatial differentiation was and remains essential to the production of a hierarchical, unitary Chinese state since 1949. It argues, specifically, that the designation of 'nationally strategic new areas' is characteristic of a longstanding and proactive attempt to retain, if not reinforce, political power through the reconfiguration of state spatiality.

Recurring Spatial Reconfigurations and the Consolidation of a Unitary State

The predominant political project in China over the last century has been to establish and consolidate a modern state structure. Pressures for state formation began during the late Qing period when the Empress Dowager, Cixi, was confronted with demands for constitutional rule. Sustained revolutionary pressures consequently triggered the demise of the Qing dynasty in 1911. In its place was a new Chinese state – the Republic of China (中华民国) – formed by Sun Yat-sen, one of the leading revolutionaries (Cohen 1988; Shambaugh 2000; Kuhn 2002). Sun's subsequent tenure was transient, lasting less than three months before his successor, Yuan Shikai, moved swiftly to re-institute the monarchy. Yuan was eventually thwarted in 1916 and fresh state-building efforts were launched by Chiang Kai-shek, the-then leader of the Nationalist Party (or Kuomintang, the KMT). During the two decades that followed, the KMT established institutions conducive to state rule such as the Ministry of Finance and Ministry of Foreign Affairs; a strong military force was also developed in tandem. The effectiveness of these institutions was undermined, however, by the KMT's inability to penetrate local communities in the rural hinterland and the total war against Japanese occupation between 1937 and 1945 (Duara 1991; Strauss 1998; Remick 2004; Osinsky 2010). It was only after the CPC emerged victorious in a protracted, three-year civil war with the KMT in 1949 that the state-building project stabilised. Andrew Walder (2015: 2) underscores the historical significance of the CPC victory:

For the first time in well over a century, there would be a Chinese state that effectively controlled its territory within secure borders, and that was able to stamp out pockets of domestic rebellion. For the first time in China's long history, salaried state officials, not local notables, would administer Chinese society in rural villages and urban neighbourhoods. These officials were part of a national hierarchy that connected the apex of power in Beijing directly, and relatively effectively, with life at the grass roots. Mao and his comrades may have viewed the victory of the Communist Party in 1949 as part of the triumph of world socialism, but it marked the birth of China's first modern national state.

Perhaps counter-intuitively, the CPC succeeded not because it formulated more innovative state building policies. As John Fitzgerald (1995) argues, state formation through the late Qing, early Republican (1912–1927), Nationalist (1928–1949) and communist periods (1949 and thereafter) was underpinned by the *continuous* quest for a unitary state in spite of differences over what kind of nation this state represents. Along the same vein, Samuel Jackson (2011: 76) points out that 'nationalist efforts to secure political unification of China during the 1920 and 1930s informed the strategies and institutions adopted by the CCP to consolidate control'. Building on these insights, this book argues that the CPC's state-building project was distinct from its predecessors because space was not construed as a passive 'container' or epiphenomenon of state-building. To paraphrase Henri Lefebvre (1991), the CPC changed China because it first altered its political-economic spatial formations.[8]

The Mao regime prioritised the integration of a unitary state through large-scale territorial reconfigurations after its victory in the Chinese civil war was imminent. This was a transformative approach because 'China' under the KMT was effectively a patchwork of disparate regional economies – dominated by warlords, Japanese colonialists and local 'land tyrants and gentry' (*tuhao lieshen* 土豪劣绅) – that significantly precluded the concentration of political power. Integrating these territories into a national whole was encumbered by poor access to local economies, particularly in the vast rural hinterland, where widespread resistance was encountered.[9] This pushback was contained and eventually broken after the CPC implemented an important but often-overlooked spatial strategy: the political–geographical division of its newly acquired territories into six administrative divisions. These regions were called North China (*Huabei* 华北), led by Liu Shaoqi; Northeast China (*Dongbei* 东北), led by Gao Gang; East China (*Huadong* 华东), led by Rao Shushi; Central and South China (*Zhongnan* 中南), led by Lin Biao; Northwest China (*Xibei* 西北), led by Peng Dehuai; and Southwest China (*Xinan* 西南), led by Deng Xiaoping. This military-styled territorialisation provided the platform for the CPC to *thoroughly* redefine rural state–society relations through the redistribution of rural land to poor peasants. Once landownership reconfigurations were completed by 1952, the CPC augmented its political power through coercive and at times violent mobilisational campaigns (Strauss 2006).

At the same time, two leaders of the six administrative regions – Gao Gang and Rao Shushi – were suddenly purged on the grounds of insubordination (Teiwes 1990, 1993, chapter 5; Shiraev and Yang 2014). Drawing from this consolidated political strength, the Mao administration rolled out plans for the mass collectivisation of means of production. This was to culminate in the Great Leap Forward industrialisation program between 1958 and 1961.

Writing in support of this economically nationalistic project, the-then Shanghai Mayor, Ke Qingshi, published an article in the Party's leading journal, *Red Flag* (*hongqi* 红旗), in February 1959 to encourage subnational actors to privilege national-level developmental goals. This article attained touchstone status in Chinese policymaking and academic circles for introducing the metaphor 'the whole country as a chessboard' (*quanguo yipanqi* 全国一盘棋). Days after its publication, the key tenets of the article were echoed and endorsed by the party-state through an editorial in its mouthpiece, *People's Daily*:

> Our socialist economy develops along planned ratios. In order to deploy enthusiasm within different spheres in the most efficient and most reasonable way, it would be necessary to enhance centralized leadership and macro arrangements, it would be necessary to look from the perspective of the whole country, and arrange the national economy in the form of a chessboard. (*People's Daily*, 24 February 1959; author's translation)

Particularly pertinent for the analysis in *On Shifting Foundations* is the editorial's insistence on placing national concerns ahead of local calculations:

> The initiatives and flexibility of every leadership organ and department during the implementation of central directives should be promoted at any time; work cannot be performed well when these characteristics are lacking. However, proactivity and flexibility should be brought forth with reference to 'the whole country as a chessboard'. These should first and foremost be used to ensure the victory of 'the whole country as a chessboard', to guarantee the actualisation of the state plan. Only when the projects and plans designated by the state are fully completed can proactivity and flexibility be extended to other areas. (*People's Daily*, 24 February 1959; author's translation)

As the subsequent chapters will elaborate, the CPC's contemporary development of 'nationally strategic' policy experimentation is premised on a repurposed 'chessboard' philosophy. The continued relevance of this philosophy is interesting because, within its original context, Ke Qingshi's imploration to subsume subnational developmental initiatives to national goals neither inspired the Great Leap Forward program to success nor enhanced economic production during the Mao era. After it became apparent that this program was unsuccessful, Mao implemented more radical measures to ensure China did not follow the 'revisionist' steps of the Soviet Union. This was first launched through the Third Front

Construction program in 1964 (*sanxian jianshe* 三线建设), on the premise of growing military threats from 'imperialism and their running dogs' (*diguozhuyi jiqizougou* 帝国主义及其走狗). Mao ordered means of production to be relocated from the coastal city-regions, officially termed the 'First Front', to those in the relatively sheltered interior, termed the 'Third Front' (ref. Li and Wu 2012: 61–64). The relocation was an immensely costly project that more negatively impacted economic recovery than the chaotic Cultural Revolution (1966–1976), as Barry Naughton (1988) shows. Once the national industrial composition was spatially reconfigured, the platform was set in 1966 for the further consolidation of CPC rule – to exterminate the 'clique of capitalist roaders' (*zouzipai* 走资派).[10]

Economic reconstruction unsurprisingly took a backseat during the first few years of the Cultural Revolution. Cell-like administrative units were tied hierarchically to Beijing in tandem with geo-economic insulation (under the nationalistic slogan of 'self-sufficiency') to ensure absolute political control. Provinces were granted significant autonomy to self-finance developmental projects in return for the enforcement of a minimal trade policy. For almost a full decade prior to the 1978 reforms, the 'Chinese economy' looked more like a customs union more than a common market; it was an entity with a common barrier against the global economy, within which free trade did not exist. Embedded within the previously mentioned 'People's Communes' and urban industrial units (*gongye danwei* 工业单位) was effectively a 'cellular' and 'fragmented' economic structure that was constituted by cell-like, self-sufficient and regionally uneven administrative units (Donnithorne 1972; Tsui 1991; Bray 2005). To be explored further in Chapter 2, these policies collectively comprised a 'politics in command' (*zhengzhi guashuai* 政治挂帅) approach to socioeconomic regulation that lasted until Mao's passing in 1976.

Chinese state spatiality was (again) reconfigured to prioritise and incentivise market-driven production under the pragmatic Deng Xiaoping leadership. After taking power in 1978, Deng immediately made it clear his priority was the survival of the CPC through the enactment of the 'Four Cardinal Principles' in the Chinese Constitution: (i) We must keep to the socialist road; (ii) We must uphold the dictatorship of the proletariat; (iii) We must uphold the leadership of the Communist Party; (iv) We must uphold Marxism–Leninism and Mao Zedong Thought. It is important to note that the third principle is the pivot on which the other three principles rest (for full text of Deng's speech on the Four Cardinal Principles, see *People's Daily*, 30 March 1979).

To fulfil his commitment, Deng's signalled his preference for Vladimir Lenin's 'New Economic Policy' in post-Tsarist Russia, which accommodated private enterprises in the quest for socialism. Interestingly, this approach marked a *return* to the CPC's original policy focus of the early 1950s (Horesh and Lim 2017). Deng's flexible approach was predicated on proactive spatial restructuring that facilitated the growing urbanisation of capital and labour power (Lin 1999; Ma 2005). Underpinning this approach was the 'ladder-step transition theory', or *tidu tuiyi lilun* (梯度推移理论). First espoused by the Shanghai-based academics Xia Yulong and Feng Zhijun in 1982, this prescriptive 'theory' attracted the attention of a senior CPC cadre, Bo Yibo, and subsequently permeated central policymaking circles. Instituted as a policy blueprint during the 7th Five-Year Plan (1986–1990), this predictive theory divided Chinese state spatiality into three economic–geographical belts: the eastern (coastal), central, and western (Xia and Feng 1982). The Deng administration allowed one belt (the eastern seaboard) the priority in ascending the development 'ladder'. It assumed that the fruits of development in the 'first mover' belt would diffuse downwards to other rungs of the ladder. This template of instituted uneven development became the *basis* of market-oriented reforms: the Deng administration expanded China's re-engagement with the global economy by permitting foreign investments beyond the first four SEZs. Prior to the 1994 fiscal overhaul that (re)concentrated fiscal resources at the central level, preferential fiscal policies were given to selected coastal provinces to accelerate their respective developments, while subsequent tax reforms continued to benefit these provinces (Wei 1996; Dabla-Norris 2005).

Up until Deng's passing in 1997, the CPC did not designate any time for accumulated capital to be proactively transferred from the coastal belt to the central and western interior to attain long-run spatial equilibrium. There was also no detailed plan that explains what would happen to the coastal provinces' economic development as resources are re-directed westwards. Ostensibly aware of potentially-damaging consequences of the widening coast–interior unevenness, Deng's successor, Jiang Zemin, began to focus on developmental issues confronting the interior provinces. Reiterating Deng's philosophy, Jiang averred in a 1999 speech that 'reducing the developmental disparities within the entire country, developing in a coordinated manner and ultimately attaining Common Affluence is a basic principle of socialism' (*People's Daily*, 10 June 1999; author's translation and emphasis). Yet, in an important qualifying point made in the same year, Jiang clarified he was not about to jettison the spatial logic of socioeconomic regulation that had contributed to China's economic growth:

> My understanding is, when Comrade Deng Xiaoping mentioned letting some regions and people to first prosper before gradually reaching Common Affluence, that is still not the end point. Upon reaching a relatively higher standard of living, more advanced regions must still move forward. Equilibrium is relative while disequilibrium is absolute, this is the objective rule of material development (Jiang 2006: 341; author's translation).

It was based on this 'objective rule' that Jiang announced, in November 1999, the 'Great Western Development' spatial project (*xibu dakaifa* 西部大开发). Ratified by the State Council in 2001, this strategic program represented the beginning of more targeted approaches towards developing the large interior regions. The original plan involved enhanced fiscal redistribution to the western provinces; more commitment by the central government to infrastructural development; opening up more sectors for foreign investments; and implementing preferential policies to attract foreign capital to the interior parts. Two other cross-provincial programs known as Northeastern Rejuvenation (*dongbei zhenxing* 东北振兴) and Rise of the Centre (*zhongbu jueqi* 中部崛起) were launched in 2002 and 2004 respectively.

While fiscal monies have been redistributed to the provinces involved in these developmental programs, the growing population concentration in the coastal cities and the huge inter-regional income disparities over the 2000s indicates the overall pattern of regional unevenness was not ameliorated. Structural change was precluded because these cross-provincial programs entailed no specific institutional (re)formulations at the provincial level (Li and Wu 2012; Liu et al. 2012). According to Weidong Liu (2012: 7), a leading economic geographer from the Chinese Academy of Sciences, the broad cross-provincial developmental programs are part of a passive overall plan; they function independently rather than relate to one another, and do not have clear goals. For this reason, Liu adds, these programs do not provide guidance for an overall (i.e. national) regional development strategy.

Unsurprisingly, as indicated previously in Table 1.1, the Chinese State Council overtly identified regional disparities and lack of overall coordination as a major policy issues in its 12th Five-Year Plan (2011–2015). In a televised press conference, the-then Premier, Wen Jiabao, emphasised that despite China's current standing as the world's second largest economy, his government is 'also fully aware that China remains a developing country with a large population, weak economic foundation and uneven development' (China Central TV live telecast, 14 March 2011; author's transcription). Shortly after, the-then President, Hu Jintao, acknowledged that 'there is lack of adequate balance, coordination or sustainability in our development' (*China Daily*, 15 April 2011). This official recognition corroborates the obdurate persistence of uneven economic-geographical development in China despite the previously-mentioned ameliorative attempts.

Against this backdrop, Xi Jinping recently implored the CPC to take 'the "whole country as a chessboard" approach in the developmental activities of the 13th Five Year Plan [2016–2020], through which coordinated development would be decisive' (*Xinhua*, 18 January 2016). Once again, this called for incorporating proactive and flexible governance within the context of the national 'chessboard'; once again, the recourse was to 'feel for stones' to overcome the new challenges. An emergent response over the last decade was to institute experimental policies in the 'nationally strategic new areas'.

Approaching the 'Nationally Strategic New Areas'

To evaluate the emergence and effects of 'nationally strategic new areas' against this backdrop of recurring spatial reconfigurations across China, *On Shifting Foundations* presents an integrated analytical framework that foregrounds the relationship between state rescaling, path-dependency and policy experimentation. This framework situates these experimental zones as integral to and an impact of the multi-dimensional process of state rescaling. State rescaling is defined as the reconfiguration of regulatory relations between the national, subnational and supranational governments, such that what represents the 'national interest' is no longer expressed and realised at one scale (i.e. nationwide). This reconfiguration is not a simple transference of regulatory capacities from a national government to governments or non-governmental entities located at other scales. The notion of 'transfer' implies a one-directional movement, whereas state rescaling is underpinned by political strategies that re-define how regulatory power is (to be) shared between governments.

This reconfiguration of political power is conceptualised as a dynamic process that shapes and is shaped by *successive* spatial divisions of regulation. First conceived by Jamie Peck (1998) and further developed by Neil Brenner (2004), this conceptualisation parallels and builds on Doreen Massey's ([1984] 1995) 'spatial divisions of labour' thesis, which views spatial change as predominantly driven by private capitalist interests. Specifically, Massey's thesis refers to the way economic investors respond to and reshape geographical inequality, with a primary focus on how the calculations of firms are affected by and in turn reshape localities. 'Spatial structures of different kinds can be viewed historically', writes Massey (1995: 118), 'as evolving in a succession in which each is superimposed upon, and combined with, the effects of the spatial structures which came before'. The key process of note is the 'combination' of different spatial structures within a broader system of production: 'The combination of layers is a form of mutual determination, of the existing characteristics of the area or regional system with those of the geographical patterns and effects of previous uses' (Massey 1995: 114–115). Massey's approach has been creatively adopted to conceptualise local(ised) expressions of regulatory shifts. Central to her approach for geographical studies of regulatory shifts, for Peck (1998: 29), is the emphasis on interaction:

> If, in a parallel sense to Massey's (1984) conceptualization of 'rounds of accumulation' unfolding unevenly across the economic landscape, this can be conceived as a process of institutional layering, then it is one which entails a considerable measure of reciprocal interaction between the layers. The process by which new geographies of governance are formed is not a pseudo geological one in which a new layer (or round of regulation) supersedes the old, to form a new institutional surface. Rather, it is a dynamic process in which national (regulatory) tendencies and local (institutional) outcomes mould one another in a dialectical fashion.

Brenner's (2004: 108) conceptualisation further foregrounds this 'dialectical fashion': 'Spatial divisions of (state) regulation are...directly analogous to spatial divisions of labour insofar as both entail determinate articulations and differentiations of particular types of social relations – whether of capitalist production or of state regulation – over an uneven territorial surface and within a chronically unstable scalar hierarchy'. Like Peck (1998, 2002), Brenner (2004: 109) calls for an understanding of the 'interactions' between new regulations and the actually-existing 'economic landscape'. While the Chinese context differed from those that informed the conceptual work of Massey, Peck and Brenner, there are striking similarities in the ways regulatory power is reconfigured through spatial restructuring. The emergence of 'nationally strategic new areas', each charged with implementing a unique set of experimental policies, would therefore be assessed in relation to the impact – both constitutive and constraining – of inherited institutions.

Institutions are defined in this book as the rules and regulations, both legalistic and informal, which structure socioeconomic interactions. This structuring process is at once a licence (it sets up the actualisation of specific actions) and a limitation (the rules set boundaries that preclude or criminalise specific actions). Propagandistic campaigns, the CPC constitution, the RMB-denominated monetary system and the so-called 'feudal' clan-based associations all constitute various aspects of formal and informal institutions in the Chinese context. The overall institutional structure is sustained and reproduced holistically through what is known within Chinese policymaking circles as *dingceng sheji* (顶层设计), literally 'top-level design'. A direct legacy of the 'chessboard philosophy' and an inversion of the Marxian base-superstructure logic,[11] this process become the *basis* that determines political-economic evolution in China. Carl Walter and Fraser Howie (2011: 8) offer an incisive description of its enduring impact:

China's economic geography is not simply based on geography. There is a parallel economy that is geographic as well as politically strategic. This is commonly referred to as the economy 'inside the system' [or *tizhinei* 体制内]...and, from the Communist Party's viewpoint, it is the real political economy. All of the state's financial, material and human resources, including the policies that have opened the country to foreign investment, have been and continued to be directed at the 'system'. Improving and strengthening it has been the goal of every reform effort undertaken by the Party since 1978. It must be remembered that the efforts of Zhu Rongji, perhaps China's greatest reformer, were aimed at strengthening the economy 'inside the system', not changing it.

This observation of continuity-through-change offers an interesting prism to assess Xi Jinping's (2013: n.p.) claim that the 'feeling for stones' approach is 'dialectically unified with top-level design'. Corresponding with Peck's previously quoted point on dialectical institutional layering, Xi's focus on 'dialectics' strongly suggests that policy experimentation is unfolding in tension with established parameters 'inside the system'. This then raises the question on the attainment of structural 'unity' at the national level: does it entail the thorough transformation, subtle refinement or further reinforcement of policies and practices associated with existing institutions?

Answering these questions entails an evaluation of the *extent* to which new initiatives are path-dependent. As existing research in urban and regional studies have shown, geographically-variegated developments are situated within historical pathways. Reviewing the connections between regional development and economic globalisation, Dennis Wei (2007: 25) notes how 'China's reform is a gradual, experiential, and path-dependent process'. At the intra-urban level, Fulong Wu (2009: 886) observes how socioeconomic inequalities 'show a strong path-dependence feature'. This corresponds with George Lin's (2007: 10, emphases-in-original) broader survey of post-Mao Chinese urbanism:

> If state–society relations have a particularly important role to play in Chinese urbanism, which in turn occupies a special position in the state scaling strategy that deals with the Chinese society, then the processes underlying Chinese urbanism would remain culturally specific and path-dependent despite the observation that certain forms of Western urbanism are replicating themselves in contemporary China under globalization.

Despite this acute awareness of path-dependency, the process has been examined largely in isolation rather than in tandem with the politics of regulatory reconfigurations and policy experimentation. Furthermore, the cross-scalar variations and effects of path-dependency are under-explored. Experimental policies can generate changes at the local scale, for instance, but may not lead to new developmental pathways at the national level. In some cases, place-specific policy experimentation could lead to the reinforcement of what Walter and Howie (2011) term 'the system'. Specifically, resistance develops at the national level because some institutions are 'locked in' to the extent that self-reinforcing effects discourage transformative change. Chapter 3 will integrate these processes within a dynamic analytical framework. In so doing, this book circumvents the challenges of state centrism, a common social-scientific approach that view states as 'containing' economies and political authority as limited within the confines of state spatiality.

Three problematic assumptions associated with state centrism will be avoided, namely spatial fetishism, methodological territorialism, and methodological nationalism. Spatial fetishism is defined as the notion that space is distinct and functions autonomously from social, economic and political processes. In contrast to spatial reification, which regards space as a thing with a materiality and life of its own, spatial fetishism treats space as a pre-given entity that needs no separate theorisation. For instance, a foundational assumption of macroeconomic theory – that economic processes could be reduced to 'aggregate demand and supply' – does not ask why it is possible or, indeed, necessary to 'aggregate' to the national scale. That the intrinsic instability of state regulation could problematise the aggregating process over time is never questioned; the assumption 'all things being equal' will not stand once the malleability of state space is considered, however. This fetishistic approach reduces locational patterns to simple cost–benefit calculations or universal laws that are neither connected to nor capable of explaining unequal power relations that are spatially (re)configured. In short, it is not concerned about how differentiated territories regulate and reproduce social, political and economic phenomena at the national scale.

Contrary to this two-dimensional view of space, research has demonstrated how the inherent instability of socio-spatial formations generates dynamic tensions with state apparatuses (Swyngedouw 1997; Peck 2002; Brenner 2004; Massey 2005; Harvey 2008). For this reason, while state space should never be taken as capable of generating causal influences autonomously, the *conjunctural specificity* of its configuration could direct or confine socioeconomic processes to the sub-national and/or supra-national scales and consequently determine whether the CPC could fulfil its regulatory objectives. This dynamic conception applies to the economic geography of 'new China', as discussed earlier in this chapter, and will be adopted to explore the rationale and ramifications of designating the 'nationally strategic new areas'.

The portrayal of the Chinese political economy as a disparate and inherently unstable territorial mosaic foregrounds two other assumptions problematised by state rescaling research: methodological territorialism and nationalism. Across much of the social sciences, Brenner (2004) observes, social processes are assumed to occur within spatially-delimited 'containers', the most privileged of which remains the nation-state. While national borders generate constitutive effects, they are also inherently porous. In view of this porosity, Brenner (2009: 48–49) emphasises the importance of *not* privileging specific scales when seeking to understand social phenomena, but to understand how and why these phenomena are scaled:

> [S]cale cannot be the 'object' of political-economic analysis, for scales exist only insofar as key political-economic processes are scale-differentiated. From this point of view, it is more appropriate to speak of scaled political economies – that is, of the scaling and rescaling of distinctive political-economic processes – rather than of a political economy of scale per se.

As previously mentioned, a key objective of this book is to foreground the *co-constitutive* relationship between Chinese state stability and regular regulatory reconfigurations. That 'China' as we know it today is an outcome of regular and in some ways cumulative state rescaling rather than a normative national 'container' illustrates methodological nationalism as the third problematic assumption. Social scientific studies predominantly collate, present and compare data at the inter-national level. While subnational socioeconomic data is often available, they are presented as subsets of national data, which underscores the constitutive effect of national borders on statistical quantification. However, these data would not be directly helpful for ascertaining the rationale and politics of designating 'nationally strategic new areas', Rather, the discourses of key decision-makers positioned at various levels of the regulatory hierarchy would be a more relevant source of information. Juxtaposing the discourses of these actors could open new ways to understand logics of regulatory reconfigurations that are overlooked by nationally-oriented quantitative data. Indeed, while the Chinese political economy comprises the site of analysis, political and economic actors have been proactively seeking to negotiate their place-specific socioeconomic interests in relation to national-level calculations through 'nationally strategic' policy experimentation.

Table 1.2 summarises how this book will address the a-spatial assumptions of state centrism. The counter-cases are delineated along the lines of geographical-historical re-assessment, theoretical re-evaluation, and case studies of 'nationally strategic new areas'. Through reappraising the spatial logics of socio-economic regulation in China after 1949, Chapter 2 foregrounds two distinct geographical processes – the centralisation-decentralisation dynamic and instituted uneven development – in developing and reinforcing the CPC's politico-economic power. This reappraisal provides the platform for challenging the assumptions of methodological territorialism and nationalism in Chapter 3. Specifically, the chapter frames the national scale as a *process*; as constituted by the *tensions* between attempts at change (policy experimentation in subnational locations) and demands for continuity (resistance to the implementation of experimental policies nationwide).

As mentioned earlier, the empirical analysis will be based on case studies about the emergence of 'nationally strategic' policy experimentation in the PRD, where Nansha New Area is designated, and Chongqing, the municipality appointed as the site of Liangjiang New Area. As is now well-documented, the PRD was the frontier of China's 'reform and liberalisation' in the early 1980s. While it remains the leading city-region in export orientation and economic output today, its developmental approach came under pressure during and after the 2008 global financial crisis, which led to a new series of experimental reforms to generate new competitive advantages for this extended metropolitan region. Prior to the designation of Nansha New Area in 2012, the CPC instituted reforms in two smaller 'new areas', one to the east (Qianhai 前海, in Shenzhen SEZ) and the other to the west (Hengqin 横琴, in Zhuhai SEZ). The reforms in both these territories were originally labelled 'nationally strategic', and were then integrated within the broader GFTZ after Nansha officially received national designation. This book will therefore examine financial and trade policies first instituted in Hengqin and Qianhai as they went further back in time (from 2009 to 2010, respectively), while also situating these policies within broader processes of economic integration in and through PRD.

Chongqing, on the other hand, became a reform test bed because it received the short end of Deng Xiaoping's developmental stick. Home to 33 million residents, the majority classified as 'agricultural', and lagging far behind coastal

Table 1.2 Refinement of state-centric assumptions through the Chinese cases.

A-spatial assumptions	Details	A refinement through the framework of state rescaling, path-dependency and policy experimentation	Counter-cases in this book
Spatial fetishism	• State space is timeless and static • State territory is unaffected by changes in the regulatory structure or broader shifts in the global system of capitalism	• Chinese state spatiality is always in the making; the current spatial form was non-existent pre-1949, and its existence is an outcome of Maoist structural and spatial reconfigurations • Current attempts at reflexive centralisation are to reinforce this Maoist politico-geographic legacy while facilitating global economic integration	• Active differentiation of state spatiality in the Mao era • Reconfiguring post-Mao regulatory spaces (Chapter 2)
Methodological Territorialism	• State territoriality is viewed as an unchanging, fixed or permanent aspect of modern statehood • The geography of state space is reduced to its territorial dimensions	• Chinese state spatiality is regularly reconfigured at different scales • Contrary to conventional portrayals, the Mao-era was characterised by entrenched uneven economic-geographical development • Post-Mao decentralised governance was and remains a dynamic extension of centralisation, not an end in itself • Spatial projects and strategies are layered on one another as well as constitute one another simultaneously • The ability of the Chinese state to produce polymorphic economic geographies within a broader global system of capitalism is the basis of contemporary Chinese statehood	• State spatiality as a dynamic process (Chapter 3) • Guangdong government's push for rescaling in response to national and global economic conditions (Chapters 4 and 5) • Rescaling in Chongqing as strategic repositioning in the global economy (Chapters 6 and 7)
Methodological nationalism	• The national scale is viewed, ontologically, as the primary and natural scale of political power and capital accumulation	• Even during the geo-economically insulated Mao-era, the national scale was determined by processes occurring at the provincial/commune and international scales • The apparent primacy of the national scale of accumulation (in relation to the global economy) is constituted by dynamic, subnational repositioning	• Explains how national-level structural coherence is intrinsically unstable (Chapters 2 and 3)

Source: State-centric assumptions adapted from Brenner (2004: 74); refinement and proposed counter-cases by author.

provinces in income and output since 1978, the sprawling city-region was chosen in 2010 to experiment with policies to overturn the (still) widening uneven development across the country. More than two decades after Deng Xiaoping pledged to reciprocate interior China's contributions to the reform process, Liangjiang New Area became the first non-coastal location to institute reforms deemed to be of national significance. These reforms build on the already high-profile attempt to overhaul the 1958 *hukou* institution in the municipality, while attempts at directing manufacturing investments to Liangjiang New Area raised questions on the relevance of coastal-oriented industrialisation (ref. discussion on Deng and Jiang's approaches to uneven development).

Although the 'national strategy' of socioeconomic development is never launched with the aim to conserve old regulatory logics, the fact that experimental policies continue to be 'contained' within specific territories reflect the difficulty of removing regulatory logics inherited from previous regimes. For instance, the previously mentioned *hukou* institution still denies social benefits to rural migrants. This has inevitably generated social discontent and created longstanding speculations about its removal (Chan and Buckingham 2008; Fan 2008). Such was the anger at this institution, 13 major newspapers took the unprecedented step of publishing a joint front-page editorial on 1 March 2010 – just before the annual 'two meetings' of top CPC delegates in Beijing – calling for its immediate removal (ref. Lim 2017). Recent reforms in Chongqing marked a tentative step in this direction, but no transformative change occurred at the national level (see Chapter 7).

The pre-existing state monopoly on financial capital supply – a Mao era legacy – similarly precludes many private investors from accessing capital from the formal financial market. While the financial system has widely adopted management mechanisms employed by market economies after the 'reforms and liberalisation' of 1978 (e.g. the public listing of banks, issuance of bonds, separation of owners from management, etc.), the entire system continues to be a function of party developmental goals (Tsai 2004; Walter and Howie 2011; Sanderson and Forsythe 2013). These phenomena jointly suggest reforms have not relinquished Mao-era regulatory logics. And it is for this reason that the contemporary reforms also reflect the constraints of institutional path-dependency.

The rationale and ramifications of the 'nationally strategic new areas' will be demonstrated through a broad spectrum of empirical data collected prior to, during and after the embarkation of three field visits to China, namely between January and February 2012 (to Beijing, with a stopover in Shanghai); in March and April 2012 (to Chongqing); and in January 2013 (to Hengqin, Qianhai, Macau and Hong Kong). A follow-up fieldtrip was made to Macau, Hong Kong and the GFTZ in June and July 2015. Given that the three primary field visits involved interactions with as many as 80 individuals on various occasions (e.g. lunches and dinner hosted by local governments, chats with guides at visitor centres of the New Areas), it was difficult to define the total number of 'interviews'.

Strictly speaking, semi-structured interviews were conducted with 31 academics and policymakers during the three field visits, while spontaneous interactions with other individuals that yielded insights into developmental processes in the 'new areas' were written into field notes. Major policy documents (e.g. the 12th Five-Year Plan; Plan for Guangdong-Macau-Hong Kong; Great Western Development plan; etc.) and published interviews by state actors were collected, translated and analysed.

Many of these documents, mostly published in Chinese and which have not been discussed by scholars outside China, offered a concrete background knowledge of the historical contexts of the 'new areas' and the reasons why these areas were given national designations. Policy analysis was complemented by an analysis of statistical and qualitative information published in the media. Close to 800 articles published in various media in China, Hong Kong, Macau and Taiwan were collated. These articles provided a significant source of information, including statistical information on updated capital flows into the 'new areas' as well as major firms that have moved into the areas. A substantial amount of this information was not reported in the statistical publications of the Chinese government, but nonetheless helped to highlight both the legacies and limitations of instituted uneven development.

Specifically, the database was constructed to illustrate and explain (i) the socioeconomic relations that constituted and are affected by the production of 'nationally strategic new areas'; and (ii) the impacts of interactions between the experimental policies and the inherited institutions in the two chosen research sites (i.e. Chongqing and the GFTZ). These two case studies helped to sidestep an important 'methodological trap', i.e. the attempt at a totalising reconstruction of the past. Through the identification of theoretically-significant empirical phenomena in the case studies (e.g. which institutions are resistant to change today, which policies the state strives to reform, etc.), new questions were generated. These questions, such as why the *hukou* institution remains so resistant to change despite the reforms to augment Chongqing's economic position through Liangjiang New Area, opened up new avenues to re-interpret events that have had specific and seemingly natural meanings attached to them. The remaining chapters will present and evaluate the implications of these new interpretations.

The Chapters Ahead

On Shifting Foundations comprises three parts that correspond with one another dynamically. The tensions between the 'chessboard' regulatory philosophy and its territorial expressions will be further examined in Part I (Chapter 2). The chapter specifically demonstrate how new rounds of socioeconomic reforms in post-1949 China, each with their distinct geographical expressions, constitute a complex palimpsest rather than a straightforward process of historical succession. Drawing on a review of published empirical evidence, the chapter complicates two dichotomous portrayals of socioeconomic 'transition' in China, namely centralisation and egalitarianism (the Mao era) and decentralisation and uneven development (the post-Mao era). It demonstrates these binaries cannot adequately explain the post-Mao economic 'miracle' when decentralised governance and uneven development also characterised the Mao era. Indeed, decentralised governance and uneven development are not antithetical to the quest for perpetual CPC rule: just as the Mao administration strategically blended centralising mechanisms with instituted uneven development to consolidate its power, the post-Mao regimes are repurposing Mao-era regulatory techniques to achieve the same objective. The chapter concludes by highlighting the need for a more incisive analytical framework that could foreground the extent to which post-Mao reforms, most recently rolled out through the designation of 'nationally strategic new areas', truly lead to shifting foundations. This challenge will be addressed in the second part of the book (Chapter 3).

Bringing into conversation research on historical institutionalism, geographical political economy and policy experimentation in post-Mao China, Chapter 3 explores how the reconfiguration of regulatory relations in China – which, as previously mentioned in this chapter, is defined conceptually as state rescaling – is dialectically entwined with inherited developmental paths established at different scales. The development of this framework is both necessary and timely because there has been a largely uncritical adaptation of the state rescaling literature, which emerged from western European and North American contexts, to the theorisation of political economic evolution in post-Mao China. This literature presupposes a relatively coherent national regulatory scale that became increasingly fragmented following an emphasis on city-regional growth. While the trend of shifting developmental resources and regulatory capacities towards city-regions may appear similar between China and advanced western economies over the last three decades, the *persistence* of Mao-era institutions in contemporary regulation underscores the necessity to probe beneath appearances (i.e. whether state rescaling is occurring or not) in order to attain more accurate understandings of the logics of rescaling (i.e. why the predominant regulatory scale has shifted to the 'nationally strategic' reform frontiers). This approach proceeds from a different point of departure: neither national-level coherence nor the movement towards urban-based accumulation is assumed as an historical inevitability.

This framework will situate the empirical research presented in the third part of this book (Chapters 4–7). It comprises two case studies of 'nationally strategic' territories in the GPRD and Chongqing. These contemporary cases are presented in two segments that each comprises two chapters (Chapters 4 and 5 on the GPRD and Chapters 6 and 7 on Chongqing). The first chapter of each segment explores how key actors built on the geo-historical context to drive the national designation of specific territories; the second examines the implications of key policy experimentation in the areas.

As Chapter 4 will show, the designations of Hengqin and Qianhai New Areas were part of a broader industrial upgrading strategy officially known as 'double relocation' (*shuang zhuanyi* 双转移) or 'emptying the cage, changing the birds' (*tenglong, huanniao* 腾笼换鸟). Rolled out by the new Guangdong provincial government just as the 2008 global financial crisis was unfolding, this new agenda aimed to relocate unwanted industries and labour power from the PRD extended metropolitan area while injecting, simultaneously, higher-order industries. The chapter shows how this strategy was contested by some actors from the central government and discursively (counter-)justified by Guangdong officials. Its over-arching objective is to show how the state rescaling process was an outcome of territorial politics between the Guangdong government, then led by Wang Yang, and senior policymakers in Beijing. Attempts by the Wang administration to introduce 'nationally strategic' sites of policy experimentation in the PRD, culminating in the 2012 designation of Nansha New Area, therefore appear as a *political necessity* to overcome the detrimental socioeconomic effects of the 'double relocation' program.

The specific reforms are documented and evaluated in Chapter 5. In Hengqin, Qianhai and Nansha New Areas, new border regulations were established to 'liberalise' flows of goods, money and people from adjacent Macau and Hong Kong (China's two Special Administrative Regions that function as open conduits to the global economy). However, a new border between Hengqin to the mainland was simultaneously constructed to prevent the 'liberalised' flow of goods and people to move smoothly into China 'proper'. Qianhai is similarly a re-bordered zone, within which approved financial institutions from Hong Kong could issue unlimited loans in Chinese yuan to Qianhai-based businesses (triggering the 'backflow' of Chinese currency from offshore centres). Nansha, the largest and most recent of the three designated territories, was developed with the intention to bring these new financial flows in connection with the 'concrete' economy. The chapter shows how the emergence of these new experimental spaces raise the question about the geographical limitations of economic liberalisation in China: Chinese policymakers want more of such 'free' spaces at the national level (hence the production of Hengqin, Qianhai and Nansha), but these spaces could only be free insofar as they are subject to new forms of geographical control.

Various developmental issues pertaining to urban–rural integration and the reduction of coastal-interior economic disparities are identified and presented in the case study of Liangjiang New Area in Chongqing (Chapters 6 and 7). To fully explain the territorial politics that led to the designation of Chongqing as a nationally-strategic reform site, a historical exploration of institutional evolution proved important. This exploration is presented in Chapter 6. The chapter discusses how the localised lack of market reforms in the Deng and Jiang eras – vis-à-vis the marketisation process along the coastal seaboard – impelled the Chongqing government to reconfigure and in turn enhance its intervention in the economy. In other words, strong state involvement in the pursuit of equitable urbanisation in Chongqing, which the *Asia Times* (24 November 2009) portrayed as emblematic of the 'resuscitation of Maoist norms', ironically did not arise as a result of some cadres' insistence on following Mao-era regulatory logics. Rather, the retention of strong state economic intervention was made possible by developmental paths generated in the post-Mao era. Interestingly, this means path-changing attempts in the post-Mao era confined some subnational economies (like Chongqing) on the old 'big state' pathway of the Mao era. As such, while the strong state in present-day Chongqing is an extension of an interventionist legacy, it now also includes post-Mao reforms as a condition of possibility. And as the chapter will elaborate, it was this strong state-directed development that enabled the Chongqing government to actualise the designation of Liangjiang New Area as a 'nationally strategic' location in the CPC's economic restructuring agenda.

Chapter 7 proceeds to evaluate the impacts of socioeconomic reforms in Chongqing. The colossal tension caused by the large-scale attempt to build public rental housing for new peasant migrants in Chongqing illustrated how two major reforms implemented at the national scale have potentially encountered limits, namely (i) the persistence of the *hukou* institution, which designated peasants in the cities as socio-spatial 'aliens', and (ii) the apparent lack of employment opportunities in the rural hinterland, which 'pushed' peasants to seek employment in the cities. The fact that equitable urbanisation is used to justify the intensification of export-oriented industrialisation (led by the Liangjiang New Area) contradicts Deng's earlier strategy to privilege capital accumulation over social welfare provision. And it is this contradiction, this chapter suggests, that led to strong opposition to the Chongqing reforms – some interest groups could have become too embedded in the path established by the Deng administration that path-changing policies had to be repudiated.

Bringing the three parts of the book together, Chapter 8 lists and critically reflects on five interrelated conclusions. These conclusions are: (i) There is much greater spontaneity and spatial selectivity in the ways initiatives of national significance are proposed, evaluated and ultimately implemented in the post-Mao era. (ii) State rescaling and geographically-targeted policy experimentation have become necessary strategies for the Chinese central government to preserve domestic socioeconomic stability vis-à-vis the limits of marketisation and an increasingly volatile global context. (iii) The main difference between the policy experimentation in the PRD (Hengqin, Qianhai and Nansha New Areas) and Chongqing (Liangjiang New Area) is how they generate new regulatory paths in the name of the 'national interest'. (iv) The policy experimentation in the two contemporary field sites is in itself filled with uncertainties and inconsistencies, which underscores the difficulties of transforming institutional foundations at the national level. (v) Designed to fit the pre-existing socioeconomic conditions of Guangdong and Chongqing, the experimental reforms are inherently contradictory: it would be unfeasible to have these experiments extended 'as is' to other locations with different socioeconomic developmental pathways. The chapter then connects these conclusions to the three-way relationship of state rescaling introduced in Chapter 3, namely the constitutive roles of place-specific develop-mental pathways; evolving national-level logics of socioeconomic regulation; and dynamic demands of transnational capital.

Read in relation to one another, the three parts of the book demonstrate how specific regulatory logics of the Mao-era have been repurposed while others were jettisoned to enable the transitional present. The institutional continuity after each round of experimental reforms since 1978 is construed as of theoretical sig-nificance: despite colossal changes engendered by new policy experimentation, the retention of inherited institutions offers a new entry point from which to investigate the relationship between the Chinese economic growth 'miracle' and Mao-era regulatory logics. The economic-geographical configurations along the eastern seaboard and the less developed interior may have changed with each round of regulatory restructuring, but, as it will be emphasised in the concluding reflections, this is engendered by interactions with the spatial logics of regulation instituted during the Mao era. To portray the post-Mao period as antithetical to the Mao era thereby precludes an exploration of inherited institutions as *capac-ities* that (seemingly still) produce a politically stable state apparatus. The impor-tant question, then, is what kinds of inherited capacities have been useful for retaining – if not reinforcing – CPC control during the post-Mao period. As the following chapters will show, contemporary policy experimentation in the 'nation-ally strategic new areas' strongly suggests it would be premature to pronounce the post-Mao political–economic 'liberalisation' as a mirror of Mao-era institutional errors or, for that matter, as a function of post-1978 reforms. Whether institu-tional changes became possible *despite* or *because of* Mao-era institutional founda-tions still needs to be established. This book takes a small step in this direction.

Endnotes

1   Full text transcribed by www.dwnews.com; author's translation.
2   Chen Yun first introduced the metaphor in April 1950. He provided its first elaborate definition in 1951: 'solutions should be reliable, this is called crossing the river by feeling for stones. Problems could emerge if we rush. We would prefer a slow and steady approach rather than be haphazard and make mistakes. This especially applies to the management of national economic problems' (Chen 1995: 152, author's translation).
3   A thorough official account of SEZ formation can be found in the memoirs of Li Lanqing (2010), the former vice Premier of China. Yeung et al. (2009) provides a detailed overview of the SEZs' evolution.
4   The designation was in part a political response to the rise of conservatism following the 1989 Tiananmen chaos. As Sang (1993) notes, Pudong New Area was allocated policies previously available only for the SEZs, which sends a strong statement on economic liberalisation given that the key centrally-governed cities (the others at the time were Beijing and Tianjin) were previously considered too important to become too open. For a detailed delineation of Pudong's emergence, see Marton and Wu (2006).

5   Chen Yun (2000: 69) made clearer his views in 1961, towards the end of the floundering Great Leap Forward program, that "On the one hand, experimentation has to be daring in thought, in talk and in doing; on the other hand, doing things specifically must be begin with actual facts, there has to be a distinction between experimentation and expansion. Expansion must involve things that have matured" (author's translation). Deng Xiaoping would mention the 'black cat, yellow cat' analogy in 1962, noting how it conveyed a Sichuanese notion of pragmatism: the colour of the cat does not matter so long as it can perform the function of catching mice.

6   The reference to 'socialism' in this book does not reflect or impose a normative conceptualisation. Following Whyte (2010), the book examines the CPC's self-proclaimed quest for socialism as an empirical fact rather than what 'socialism' means when it is measured against a particular template (e.g. Marx's version based on the experience of western capitalist economies or Lenin's version based on the largely agrarian Russian economy). While the CPC claims both Marx's and Lenin's versions to be relevant to its quest (at least ideologically), its official commitment to creating 'socialism with Chinese characteristics' is path-setting given the collapse of the socialist internationalist movement. What 'socialism' means for the CPC is thereby an empirical question, to be explored in tandem with marketisation reforms and the emergence of new planning capacities in the party-state apparatus.

7   This geographically nuanced approach is aligned with Aihwa Ong's (2004, 2006) works on graduated sovereignty and spaces of neoliberal exceptionalism in China; Douglas Zeng's (2010) edited collection on China's special economic zones and industrial clusters; and Sandro Mezzadra and Brett Neilson's (2013) analysis of multiple internal boundaries within China.

8   The original expression from Henri Lefebvre (1991: 190), a major social theorist of spatiality, was 'to change life [...] we must first change space'.

9   Despite an expanding industrial base between 1912 and 1949, as expertly demonstrated in Chang (1969[2010]), there was a wide array of economies that were not regulated by a unified state regime (Meisner 1999). Many of these economies were controlled by warlords, and this made it difficult for the KMT to centralise power (see Jackson 2011).

10   For an extensive analysis of the specifics of the Cultural Revolution, see Dikötter (2016). The rationale of the Cultural Revolution is set within a broader context of de-Stalinisation in Walder (2015).

11   The 'superstructure' in Marxian terms refers to social aspects such as culture, ideology and religion. The 'base' refers to the means and social relations of economic production. Economic production in this regard refers to the creation of things needed by society. Marxian logic states the economic 'base' generates the 'superstructure', a logic turned on its head in the latter half of Maoist rule. Distinguishing his approach from that of Stalin, Mao argued in the late 1960s that the Stalinist regime 'speak only of the production relations, not of the superstructure nor politics, nor the role of the people' (Mao 1977: 136). Through the Cultural Revolution (which officially lasted between 1966 and 1976), Mao went on to prioritise ideological purity over economic production. It was arguably because of this that the Soviets subsequently charged the CPC of moving from the fundamental tenets of Marxism-Leninism towards a new path of Maoist voluntarism.

Part I
A Geographical–Historical Re-appraisal

Chapter Two
Chinese State Spatiality as a Complex Palimpsest

In a society in which the government dominates all spheres of social life, as in China's centrally planned socialist system before 1978, it is…presumed that social equality would as a matter of course be promoted very systematically…However, whether planners and other bureaucrats adopt and implement policies that foster equality or generate inequality depends on their goals, priorities and perceptions of societal needs… So specifying the role of the socialist state in counteracting or aggravating any particular inequality is an empirical question, not something that can be assumed almost by definition (socialist = equality).
– Martin Whyte (2010: 4–5)

Everyone is talking about deepening reforms, but what is to be reformed? What are the problems that reforms in China are addressing? Of course many people are of the opinion that we were a planned economy in the past and right now we have become a market economy, but is it really that simple? – Wen Tiejun, senior economist and long-time consultant to the Communist Party of China.
(Interview with Nanfang Ribao, *23 June 2013; author's translation).*

Introduction

During a meeting with the Danish Premier, Poul Hartling, in October 1974, the Chairman of the Communist Part of China (CPC), Mao Zedong, observed that approaches to socioeconomic regulation in post-1949 'new China' were 'not

On Shifting Foundations: State Rescaling, Policy Experimentation and Economic Restructuring in Post-1949 China, First Edition. Kean Fan Lim.

much different from the old society' (CCCPC Party Literature Research Office 1998: 413). Shaped by feudalism, warlordism and imperialism, regulatory approaches in the 'old society' comprised the wage system, the allocation of rewards according to expended labour power, and the exchange of commodities through fiat money (mainly the *fabi* 法币 and Gold Yuan issued by the Nationalist Party and the yen used by Japanese colonialists in Manchuria). At the same time, however, Mao was highlighting an institutional invention he considered definitive of 'new China': the collective ownership of land, labour power and money.[1] This juxtaposition of nationalised means of production with quintessentially Marxian categories of 'capitalistic' processes explicitly – if inadvertently – underscored a foundational aspect of socialist 'new China': to develop a national economic project through the extraction of monetarily-defined surplus value from labour power. While much has changed in the four decades after the Mao regime, the transposition of the CPC into a more proactive player in the global system of capitalism arguably gives Mao's observation a contemporary resonance.

The seemingly paradoxical entwinement of state-driven capital accumulation with market oriented reforms in the post-Mao era is exemplified by the prominence given to the methods of the 'old society' during the Third Plenum of the 18th Party Congress. Chaired by the newly installed Xi Jinping administration in November 2013, the Congress proclaimed 'decisive' roles for market mechanisms in socioeconomic regulation. Concurrently, however, it pledged to 'persist with the dominant role of the public ownership system, give rein to the leading role of the State-owned economy, incessantly strengthen the vitality, strength of control and the influence of the State-owned economy' (Communiqué of the Third Plenum, published in *Xinhua*, 12 November 2013b; author's translation). To Xiao Yaqing, Director of the State-owned Assets Supervision and Administration Commission of the State Council (SASAC), this persistence is necessary because state-owned enterprises (SOEs) are led by CPC cadres, 'the most solid and reliable class foundation' (*South China Morning Post*, 17 June 2017b).[2] In a recent demonstration of its commitment to align market mechanisms to political objectives, the Chinese central government expended an estimated 5 trillion *yuan* (~US$800 billion) in the summer of 2015 on a military-styled intervention to rescue domestic stock markets from a full-blown crisis (*Caixin*, 16 July 2015; *Reuters*, 23 July 2015). These developments jointly exemplify a distinct trend in post-Mao China: change in the form of marketisation and global economic integration remains entwined with regulatory objectives first instituted during the Mao era.

This chapter appraises this process of continuity-in-change through a geographical perspective. It demonstrates how an increasingly open Chinese political economy was facilitated by two Mao-era logics of spatial regulation, namely (i) the dynamic entwinement between centralisation and decentralisation and (ii) and instituted uneven development. In so doing, the chapter illustrates and complicates two temporally-defined binaries, namely the characterisation of 'centralised' socioeconomic regulation during the Mao era and decentralised capital accumulation during the post-Mao era, and the portrayal of 'social and spatial inequalities', to re-borrow Heilmann and Perry's (2011: 10) terms, as outcomes of post-Mao governance vis-à-vis more egalitarian developmental strategies under the Mao administration. Political–economic centralisation in the late 1950s – primarily instituted through the 'scaling up' of household-based land-ownership to the communes and the state, the rural collectivisation of economic production and the strict control of demographic mobility – was *simultaneously* premised on decentralisation and instituted uneven development. In fact, urban–rural inequality peaked during the 1958–1960 Great Leap Forward in the history of 'new China'; inter-provincial disparities were pronounced throughout the Mao era; and exchange within provinces relied more on multiple currencies in the form of ration tickets than the centrally-issued *renminbi* (RMB). These develop-ments collectively suggest new rounds of socioeconomic reforms in 'new China', each with their distinct geographical expressions, constitute a complex palimpsest rather than a straightforward process of historical succession. And it is against this multi-layered backdrop that the emergence and effects of 'nationally strategic new areas' are analysed subsequently in this book.

The discussion is organised in three parts. The next section will evaluate the centralisation–decentralisation relationship following the establishment of 'new China'. This is followed by a summary and evaluation of empirical data on uneven development during and after the Mao era. These sections jointly offer a more dynamic theorisation of Chinese politico-economic evolution since 1978: it fore-grounds specifically the interactions between inherited institutions and reformist initiatives to enhance national-level regulation. The chapter then concludes by considering how decentralised governance and uneven development are not antitheses of CPC rule; just as the Mao administration strategically blended cen-tralising mechanisms and subnational autonomy with instituted uneven development to consolidate its power, the post-Mao regimes are repurposing Mao-era regulatory techniques, expressed most recently in the designation of 'nationally strategic new areas', to achieve the same objective.

The Centralisation–Decentralisation Entwinement

The rapid economic growth in post-Mao China is widely construed as driven by decentralised local governments. 'Local' (*difang* 地方) is broadly used in China with reference to administrative levels that are not 'central' (*zhongyang* 中央), namely the provincial, sub-provincial, prefecture, county, etc. 'Whereas the central state set the reform process in motion and provided localities with the incentives and the leeway to develop economically', writes Jean Oi (1992: 101), 'it is local government that has determined the outcome of reform in China'. Gabriella Montinola et al. view post-Mao reforms as characteristic of a form of 'market-preserving federalism', a strong claim at the time and even today given the unitary political system that emphasises absolute power in the central government (cf. Shirk 1993; Montinola et al. 1995; Huang 1996; Lawrence and Martin 2013). To Yingyi Qian and Gérard Roland (1998: 1156), 'one of the most distinct features of China's transition has been associated with devolution of authority from the central to local levels of government'. This observation builds on an earlier claim by Qian and Barry Weingast (1995: 13) that the 'critical component of China's market-oriented reform, which began in 1979, is decentralisation'.

Decentralised regulation in the post-Mao era has broadly been defined as a three-pronged process. It comprises (i) new policymaking autonomy given to local governments; (ii) new fiscal responsibilities to finance socioeconomic projects that were either previously undertaken by the central government or were engendered by new 'supply-side' socioeconomic policies; and (iii) direct responses by local governments to the competitive pressures generated by economic integration into the global system of capitalism (see, inter alia, Chung 2000; Su and Yang 2000; Remick 2004; He and Wu 2009; Tsing 2010). The common approach in these works is the attribution of the post-1978 economic growth 'miracle' to the devolution of regulatory capacities to subnational entities (primarily cities and rural townships). In other words, there is a positive causal relationship between decentralised governance and rapid economic growth.

These empirical accounts have illuminated received knowledge of the fast-changing Chinese political economy in two major ways. First, they demonstrate cogently how the Chinese central government does not – or, indeed, cannot – function as an omnipotent and omnipresent allocative-cum-redistributive institution. This is because of the inherent difficulty for central planning agencies to gain timely access to information on demand and supply across what has always been a geographically expansive and socially heterogeneous economy. Indeed, centralised control during the Mao-era existed only under strict control of personal freedoms (Greenhalgh and Winckler 2005; Dikötter 2010; Yang 2012). And even so, commodity supply never prioritised swift responses to (domestic) consumer demand. Second, the proactive approach to reconfigure Chinese state space in relation to the demands of economic liberalisation underscores the territory-based politics intrinsic in central governance (cf. Howell 2006; Wei 2007; Peck and Zhang 2013). This is exemplified through the Deng administration's strategy to cater to the interests of provincial officials – or 'playing to the provinces' – in order to develop potent counterweights against senior conservatives in Beijing (Shirk 1993). Underpinning this approach was a centralising logic: the reformers could retain the regulatory foundations of the Maoist state and facilitate a smoother transition to market-like rule through devolving greater authority and benefits (*fangquan rangli* 放权让利) (Shevchenko 2004; Chien 2010).

Studies have shown how increased autonomy in provincial decision-making unleashed GDP growth through stretching centrally defined parameters. Adam

Segal and Eric Thun (2001) found centrally instituted economic policies often encounter re-interpretations and/or resistances at the subnational scales. Andrew Wedeman (2001) similarly explains how the institutional structure of the CPC encourages 'strategic disobedience' at local levels. This dynamic response by local governments triggered observations of the rise of – or, more accurately, a *return* to – 'mountain-stronghold mentalities' (*shantou zhuyi* 山头主义) or 'economic feudalism' (*jingji zhuhou* 经济诸侯). Jean Oi (1992: 126), for instance, predicted 'the success of local state corporatism may in the long run force the emergence of something akin to a federal system that more clearly recognises the rights and power of localities' (see also Oi 1998). While a federal system did not emerge, local governments went on to expand their 'rights and power' to drive growth projects. 'During the implementation of the central government's macro adjustment policies (*hongguan tiaokong* 宏观调控)', reflects Han Baojiang of the Party School of the Central Committee of the CPC:

> An 'excited phenomenon' amongst local governments, which shouldn't have occurred, became apparent: orders were disobeyed and prohibitions defied (*lingbuxing, jinbuzhi* 令不行,禁不止), there was no adherence to the adjustment discipline, the acceptance of orders were only superficial (*yangfengyinwei* 阳奉阴违)...People were even joking that, during the course of regional development in China, a weird phenomenon has occurred: the more a local government covertly disagrees with the central government's macro adjustment policies and behaves more opportunistically, the more the local economy benefits; vice versa. (Han, interview with *China Economic Weekly*, 2006; author's translation)

Han's observation corresponds with empirical research that portrays growing tensions within central-local dynamics. Transgressions within subnational agendas against macro-level policies are evident in at least three domains where the central government tried to exert more control, namely to reduce (i) investments in pollutive industries (Economy 2011; Lan et al. 2012); (ii) clandestine financial transactions through the 'shadow banking system' (Tsai 2004; Li and Hsu 2012; Breslin 2014); and (iii) speculative investments in real estate (Huang and Yang 1996; Guo and Huang 2010). Emergent debates over the superiority of regional developmental approaches further suggests bottom–up constraints on national economic integration have (re)surfaced (Qiu 2011; Qu 2012; Zhang and Peck 2016; ref. discussion on local protectionism in Section 2.4). Vis-à-vis the recent injunctions by the central government to reduce GDP growth targets and correspondingly rein in the powers of prefecture- and county-level urban governments, recurring claims of positive correlations between subnational regulatory autonomy and post-Mao GDP growth are certainly not without credence.[3]

A series of historically-engaged reflections paint a more fine-shaded picture of decentralisation, however. As Yi Li and Fulong Wu (2012: 92) put it, 'decentralisation is a state strategy', and Wu Jinglian, a senior economic consultant to the CPC, offers a vivid recollection of how this 'state strategy' was, in a contradictory twist, entrenched in the two decades prior to the 1978 reforms:

> Following the decentralisation of administrative power in 1958 and 1970, chaotic scenes emerged in the economy. Where did the problem come from? Those decentralising initiatives were implemented within the framework of a 'commandeering economy'. While the microeconomic decisions in enterprises remained totally subject to hierarchical control, day-to-day regulatory power was decentralised. Apart from defence industries and some industries of an experimental nature, all enterprises experienced decentralisation to very low levels. From production planning, fixed capital investment, redistribution of goods and capital, financial and fiscal management, credit provision, all processes were decentralised. (Wu, in Wu and Ma 2012: 49; author's translation)

As Wu elaborates, decentralisation led to what was introduced in Chapter 1 as a 'cellular' economic structure:

> The outcome was a plethora of 'centres', every local government sought not to fall behind by comparing and amending planning targets, [as such] the targets kept increasing, all sorts of 'satellites' [the nickname of the time for ambitious targets, termed after the Soviet launch of the Sputnik satellite that Mao strongly admired] were released this way. (Wu, in Wu and Ma 2012: 49; author's translation).

Within this 'commandeering economy' (*mingling jingji* 命令经济), orders were issued randomly without much prior expert knowledge or relation to the national 'plan'. Such randomness – termed 'wanton guidance' or *xia zhihui* (瞎指挥) – was particularly pronounced during the Great Leap Forward and the Cultural Revolution and contrasted the scientific rationality that enabled the Soviet Union to determine the production and exchange of a vast range of commodities (Qin 2004: 77–78). Not only did 'irrational' commandeering undermine the 'chessboard' regulatory approach that was outlined in Chapter 1, it generated social chaos, widespread poverty in the rural hinterland and entrenched uneven development across the country. For this reason, Wen Tiejun, another senior economist and a long-time consultant to the CPC, offers a thought-provoking contention that the Mao era could not even be reduced to the umbrella concept of a 'planned economy' (*jihua jingji* 计划经济):

> If you seriously conduct a little research on contemporary economic history, you will discover before the reforms in 1980, China did not even get to launch a planned economy, or there was never an attempt to implement a planned economy in the true sense of its meaning. For us to come up with what resembled a plan and then make the plan reality, this only occurred after the Cultural Revolution, after the chaos was alleviated. It was only from the 1980s that China had the conditions

to devise a plan, to implement a plan, what came before could be considered a wartime economy, those were exceptional circumstances and could not count as a planned economy. (Wen Tiejun, in *Nanfang Ribao*, 23 June 2013; author's translation)

If Wen is correct, this 'irrational' regulatory logic enabled the CPC to consolidate its control over the newly formed political economy through a systematic reconfiguration of the means and social relations of production. Private capitalists and the 'landlord and gentry tyrants' (*tuhao lieshen* 土豪劣绅) were exterminated in the process, their assets transferred in turn to large state-owned enterprises (SOEs) and/or rural collectives. The previously dynamic financial system was totally nationalised (see Box 2.1). To understand how this was possible, it would be necessary to discern the role – or, as Whyte's (2010) terms quoted at the beginning of this chapter show, the primary intentions – of the socialist state under Mao.

During its first decade of power, the Mao government wanted to construct a socialist economy modelled after a Stalin-styled Soviet Union (Li 2006; Bernstein and Li 2010). The antithesis to this objective process was individual production (*dan gan* 单干), a lethal capitalistic seed that had no room in Mao's ultimate vision of Chinese socialism. To actualise this vision, the decentralisation of 1958 occurred simultaneously with the enhancement of centralised CPC control over bureaucratic functions. Political power was concentrated in the Party Committee, which was then funnelled upwards to the General Secretary (the most powerful position in the CPC technically). In an agenda published in January 1958 termed '60 Working Principles', Mao Zedong (1999a: 345; author's translation) proclaimed 'major power would be monopolised, minor authority decentralised; the Party Committee would decide, everyone else acts' (*daquan dulan, xiaoquan fensan; dangwei jueding, gefang quban* 大权独揽,小权分散;党委决定,各方去办). What followed in June 1958 was the establishment of five bodies overseeing finance, legislation, foreign affairs, science, and culture. Mao clarified in an accompanying directive that:

> These agencies all belong to the central party and are directly under the purview of the Politburo (*zhengzhijü* 政治局) and the Secretariat (*mishuchu* 秘书处). The overall political agenda will be set by the Politburo, the specific implementation by the Party Secretariat. There is only one political design arena, not two. Both agenda setting and implementation are integrated, there is no distinction between party and politics. (Mao 1992: 268; author's translation)

With this injunction, the CPC effectively hollowed out the State Council. The state apparatus of 'new China', technically an independent structure, was subsumed under the party apparatus to become a party-state. This meant the state bureaucracy became a subservient function of party calculations and interests, a phenomenon that persists in the present across literally all socioeconomic spheres despite subsequent rounds of reforms (see case studies in Chapters 4 and 6).

The Mao administration then concentrated its efforts in ideological purification, or in official Marxian parlance, the enhancement of the previously-mentioned 'superstructure', after integrating the state within the party. Highly decentralised territorial units were left to drive the economic 'base'.

Most prominent of these units was arguably the People's Communes (*renmin gongshe* 人民公社). In August 1958, the Chinese central government issued the 'Resolution to establish People's Communes in rural villages'. The official injunction called for four actions on the part of cadres, namely, (i) To ensure the merger of politics and commune (*zhengsheheyi* 政社合一) through the integration of workers, farmers, soldiers, students and merchants; (ii) Emphasise the 'bottom–up' approach to commune formation, with smaller units coming together to select the management committee of the bigger commune; (iii) In the process of merging cooperatives, the 'spirit of communism' should be applied where addressing differences in the assets and debts of each pre-existing cooperative; (iv) Indicate the current ownership institution of the people's commune is collectivised ownership, with the possibility of full ownership by the people in the future, as preparation for the transition to communism.[4] Following this injunction, 740,000 advanced cooperatives were amalgamated into 26,000-odd People's Communes. Almost at the same time, the Great Leap Forward industrialisation strategy – powered by rural collectivisation rather than the urbanisation of capital – was launched.[5]

This ambitious strategy was unsuccessful, however. What ensued was mass starvation, with fatalities estimated at between 30 and 40 million (Chang and Halliday 2007; Dikötter 2010). Senior CPC cadres like Liu Shaoqi, Deng Xiaoping, Li Fuchun, Chen Yun and Deng Zihui responded by revisiting earlier plans to facilitate individual household production and allow households to retain greater shares of their output (*baochan daohu* 包产到户). Fundamental to this down-scaling proposal was flexibility – the goal was to increase output and alleviate the deleterious socioeconomic effects of the Great Leap Forward. It was at this point in 1962 that Deng Xiaoping first presented the flexible 'yellow cat, black cat' outlook that subsequently undergirded the post-1978 reforms:

> Comrade Liu Bocheng often uses a Sichuan phrase 'Yellow cat, black cat, whichever catches the mice is a good cat'. This refers to fighting battles, that we were able to defeat Chiang Kai-shek was because we did not play by old rules, we did not follow old ways of fighting, everything depended on circumstance, what counted was victory. To recover rural production currently also depends on circumstance, that is, the relations of production cannot adopt a fixed and unchanging form, whichever form stimulates the enthusiasm of the masses shall be adopted. (Deng speaking on 7 July 1962; published in Deng 1994a: 323; author's translation)

Mao would have none of any 'old rules', however. For this and other 'right leaning' (*youqing* 右倾) proposals, Liu and Deng were castigated as 'capitalist roaders' (*zouzipai* 走资派) at the onset of the Cultural Revolution in 1966. The CPC's hard-handed negation of household-based production in the 1960s illustrates two interconnected regulatory logics. First, what the Mao administration wanted was a particular approach to capital accumulation, namely one led by rural collectivisation and the systematic transfer of surplus value to industrial projects to fulfil specific economically nationalistic goals (i.e. surpassing the UK and US and ultimately matching the Soviet Union).[6] Second, the 'relations of production' integral to Deng Xiaoping's concerns had to be premised on and reproduced through a particular territorial form. This territorial form was the People's Communes.

The symbolic and practical values of the Communes are clear. Symbolically, the CPC not only wanted to emulate communes established across the Soviet Union, but to supersede Soviet collectivisation. As Mao acknowledges in a 1957 speech: 'We are now developing manufacturing and strengthening agricultural industries. We are more courageous than Stalin by launching industrialisation within the Communes' (cited in Deng 1998: 44; author's translation). Practically, it was more manageable for the National Planning Commission, the-then central planning agency in charge of the highly complex agricultural production system, to transmit information to 26,000 communes than if it was done at the next administrative level, the previously-mentioned production brigades (*shengchan dadui* 生产大队) (Jones and Poleman 1962; Perkins and Yusuf 1984). As Box 2.1 elaborates, this territorial reorganisation formed the basis for financial centralisation during the Mao era. Perhaps more importantly, subsequent financial reforms took place under the same territorial structure despite the dissolution of the communes in 1984. Taken together, these developments underscore how the re-territorialisation of 'new China' became a precondition and an outcome of state-driven capital accumulation (cf. Shevchenko 2004; Lin 2011; Horesh and Lim 2017).

The concurrent occurrence of political centralisation with territorial and economic decentralisation since 1949 calls into question the temporal decoupling of the centralisation-decentralisation relationship. This chapter foregrounds two issues. First, the years 1978–1979 is construed as a historical watershed that distinguishes the transposition from a centralised era (driven by a powerful central bureaucracy) to a 'transitional' period that is decentralised (and indicative of a deepening of market-like rule). What has since changed is apparently a one-track scalar devolution from a centrally planned economic production to variegated territories driven by 'entrepreneurial' local governments. This book will demonstrate, however, that post-Mao socioeconomic regulation cannot be characterised as a linear scalar shift from the national scale to smaller territorial units. Rather, territorial reconfigurations in the post-Mao era exemplified a *qualitative* shift from one decentralised 'cellular' structure – to re-borrow Donnithorne's (1972) conceptualisation of the Mao-era economy, as highlighted in Chapter 1 – to another based on city-regional capital accumulation, provincial-level protectionism and integrative national-level regulation

---

Box 2.1   Financial (De)centralisation in China Since 1949

The Centralisation–Decentralisation Entwinement within the Chinese Financial System, 1949–Present

Arguably one of the definitive economic changes in the post-Mao era was the reform of the Chinese financial system. In some ways, these 'reforms' were more aptly termed 'experimental additions' of actually-existing institutions common in many countries, and were arguably necessary to facilitate a transition from an insulated economy to one that would be increasingly integrated with the global system of capitalism.

After the CPC established 'new China' in 1949, all of the pre-1949 capitalist companies and institutions were nationalised by 1950. Between 1950 and 1978, China's financial system consisted of a single bank – the People's Bank of China (PBoC), a central government owned and controlled bank under the Ministry of Finance, which served as both the central bank and a commercial bank, controlling all the formal financial assets of the country and handling almost all financial transactions. With its main role to finance the physical production plans, the PBoC used both a 'cash-plan' and a 'credit-plan' to control the cash flows in consumer markets and transfer flows between branches. All banking facilities in the communes and cities were connected to the sprawling PBoC control network. If anything, financial nationalisation was an exemplar of Mao-era political-economic centralisation.

Yet, the communes were decentralised financial units at the same time. Central to this decentralised function was the predominant use of 'ration tickets' as daily currency. As Jan Prybala (1967: 175) estimates, 'in the commune at least half the income was distributed in the form of ration tickets redeemable at the commune mass halls and tailoring establishments' (for a detailed overview, see Li 2016).

The first main structural change commenced in 1978 and concluded in 1984. By the end of 1979, the PBOC departed the Ministry and became a separate entity, while three state-owned banks took over some of its commercial banking businesses: The Bank of China (BOC) was tasked to specialise in transactions related to foreign trade and investment; the People's Construction Bank of China (PCBC), originally formed in 1954, was established to handle transactions related to fixed investment (especially in manufacturing); the Agriculture Bank of China (ABC) was set up (in 1979) to deal with all banking business in rural areas; and, the PBoC was formally established as China's central bank. This marked the beginning of a two-tier banking system in China. Finally, the fourth state-owned commercial bank, the Industrial and Commercial Bank of China (ICBC), was formed in 1984, and took over the rest of the commercial transactions of the PBOC.

Throughout the 1980s, the development of the financial system can be characterised by the fast growth of financial intermediaries outside of the 'Big Four' banks. Regional banks (partially owned by local governments) were formed in the Special Economic Zones (SEZs) in the coastal areas; in rural areas, a network of Rural Credit Cooperatives (RCCs; similar to credit unions in the U.S.) was set up under the supervision of the ABC, while Urban Credit Cooperatives (UCCs), counterparts of the RCCs in the urban areas, were also founded. Non-bank financial intermediaries, such as the Trust and Investment Corporations (TICs; operating in selected banking and non-banking services with restrictions on both deposits and loans), emerged and proliferated in this period. This period also marked the re-emergence of what is now (in)famously known as the 'shadow banking system', with possibilities for private, unregulated lending emerging in tandem with reforms in the rural hinterland (Tsai 2004).

The most significant event for China's financial system in the 1990s was the inception and growth of China's stock market. Two domestic stock exchanges (in Shanghai and Shenzhen) were established in 1990 and grew very fast during most of the 1990s and in recent years in terms of the size and trading volume. In parallel with the development of the stock market, the real estate market also went from non-existent in the early 1990s to one that is currently comparable in size with the stock market. Both the stock and real estate markets have experienced several major corrections during the past decade, and are characterised by high volatilities and speculative short-term behaviours by many investors.

The financial mechanisms found in most capitalist economies today are employed by the CPC in financial regulation. In an intriguing twist, however, the new system that emerged in the post-Mao era is beginning to take the form of the PBoC-styled control network employed in the Mao era: the entire system trades – and, by implication, distributes risks – almost within itself (see Walter and Howie 2011). The dominant banks are now owned by the central government; regional banks by local governments. All banks are subject not simply to moral suasion by the PBoC; mandate-like directives constitute the standard governance method. Correspondingly, lending to private SMEs is a major issue, with banks preferring to lend to larger firms that are invariably more likely to be state-linked or state-owned. As Tsai (2004: 7) shows in an excellent analysis, the market-like transformation plays an important role of taking over the burden of subsidising SOEs as the central government undertook fiscal reforms. This development ultimately limited 'the availability of formal bank credit to the most productive and market-oriented part of the Chinese economy, the private sector'.

The persistence of strong state control is an important phenomenon in itself and has been the focus of scholarly research in recent years. It takes on a different significance in the context of research on 'nationally strategic new areas': how, indeed, could the ongoing attempt to 'internationalise' the Chinese *yuan* take place without the CPC relinquishing control of a financial system it painstakingly reconfigured to reinforce state dominance? The experimental reforms in Hengqin and Qianhai New Areas may yield some answers (see Chapter 5).

(ref. Chapter 3, second section). As the remaining chapters of the book will elaborate, this shift is most recently exemplified through the designation of 'nationally strategic new areas' after the 2008 global financial crisis.

Connected to this shift is the second issue: taken as an endogenous phenomenon independent of political centralisation, 'decentralisation' takes on the appearance of a self-contained process that shaped the 'miraculous' outcomes of post-1978 economic reforms in/across China. While there is no doubt that some subnational governments in China have implemented creative socioeconomic policies that produced positive results following the delegation of more administrative power, the causal logic of policy innovations need not be related to decentralisation in a linear-sequential fashion (i.e. decision to decentralise → innovative local policies of/for capital accumulation → positive local socioeconomic outcomes). Indeed, this linear-sequential logic is unclear in three ways. First, it does not explain whether the strong economic growth in some regions is really and/or solely due to local policy innovation. Second, it does not explain why some local governments are more innovative than others. Third, this logic suggests China's post-1978 economic restructuring is diverging from a top-down mode of regulation when, as Box 2.1 demonstrates, this mode has been repurposed to align with the demands of the global economy. The next section will elaborate how the legacy of decentralised governance triggered new rounds of centralising measures in the 1990s as well as consolidated the importance of the urban–rural 'dual structure', a socio-spatial regulatory tool instituted in 1958. Furthermore, assessing whether 'changes' to central-provincial interactions are 'genuine' is challenging because an 'objective' yardstick is required to ascertain authenticity. What constitutes objectivity is intrinsically political, however.

Bearing these issues in mind, this book is careful not to reify decentralised governance – expressed most recently through the 'scaling down' of 'nationally strategic' policy experimentation in different urban territories – as an autonomous driver of post-Mao socioeconomic reforms. As Hongbin Cai and Daniel Treisman (2006: 506) put it, the key reforms that 'reshaped China's economy began in the late 1970s and early 1980s, before any significant decentralisation had occurred. In fact, China's authoritarian centralisation helped speed the geographical spread of policies found to work well'. Along the same vein, Laurence Ma (2005: 478) observes the rescaling of China's nation-building efforts downward in the post-1978 reform era represents not so much the retreat and disarticulation of the central state as a rearticulation of state power with a different form of state intervention at lower spatial scales. These findings correspond with Li's (1997: 49) proposal that the central-provincial relationship in China must be conceptualised as a 'non-zero-sum, interactive process of conflicts and compromises, whereby genuine changes to the relationship are made'. Building on these studies, this chapter argues that a more fruitful research avenue – which was pursued in this book through the case studies of Chongqing and the PRD – is to examine *whether the territorial reconfiguration of regulatory policies have altered the fundamental roles and objectives of the CPC.*

Uneven Economic–Geographical Development: Undesirable Outcome or Developmental Precondition?

The evolution of 'new China' was and remains characterised by uneven economic–geographical development. Empirical studies consistently reveal entrenched income inequality between the urban and the rural meso-scales and between provinces (cf. earlier reviews by Fan [1995] and Wei [2007]). Donnithorne (1972: 618) notes how 'efficient cells do not subsidize the inefficient' following Mao's down-scaling of 'self-reliance' to individual economic units during the Cultural Revolution, a development that 'of course militates against egalitarianism' (cf. Snead 1975). Interprovincial income disparities did not narrow between 1952 and 1985, as Kai Yuen Tsui (1991) discovers. This corresponds with Thomas Lyons' (1991) analysis of output data between 1952 and 1987. Instead of reducing income inequality, Lyons (1991: 499) found 'the absolute gaps between richest and poorest widened considerably between the 1950s and the mid-1980s'. For this reason, Lyons (1991: 499) concludes that the Mao era 'cannot be viewed as a wholly successful implementation of the egalitarian ideals widely thought to underlie the Chinese model of development'.

A separate study on real per capita consumption between 1952 and 2000 by Ravi Kanbur and Xiaobo Zhang (2005) reinforces the validity of these studies. They identified sharp growth in inter-provincial inequality (i) in the buildup to the Great Leap Forward (1955–1960); (ii) after the Cultural Revolution began in 1966; and (iii) after the Deng administration made a distinct turn to urban-oriented industrialisation in the early 1990s. The key finding of this study was the return of urban–rural inequality in 2000 – albeit at higher levels of income – to that of 1978 (ref. statistics collated in Figure 3.1; cf. Dunford and Li 2010). The similarity of these levels, one before the reforms and one two decades after, is empirically significant because it indicates the exacerbation of inequality was not solely an outcome of marketisation. By implication, this calls for a qualitative exploration of factors that contributed to uneven development in post-Mao China.

Research indicates enduring uneven development is entwined with the 'scaling up' of the national fiscal system during the 1990s. As mentioned in the previous section, power centralisation during the Mao era was actualised through decentralised governance, and this strategy severely undermined the financial power of the central government. Tsui (1991) suggests that the connection originated from the launch of the Great Leap Forward industrialisation campaign in 1958. The decentralised fiscal governance that accompanied this campaign witnessed the doubling of the share of extra-budgetary revenue, defined as fiscal funds not subject to control or extraction by the central government, in the 1960s. As Tsui (1991: 15) puts it, this trend went on to exhibit 'a distinct upward trend in the 1970 and 1980s'. The ability to raise funds independent of the central government directly contributed to unevenness in economic output and standards of living during the Mao era. Wen

Tiejun, the long-time policy consultant to the CPC, corroborated this point in an interview with *Nanfang Ribao* (23 June 2013; author's translation):

> What is known as localised development began in 1957, when the Soviet Union stopped giving China aid related investments. The central government's budget fell, this budget became a 'red line' while local budgets were the 'blue lines'. Because there was no more money [in Beijing], there was no true centralisation of power in China since the 1950s. At the time, all the older comrades knew of this phrase 'central finance is sitting on the slides, local finances are soaring through the skies' [*zhongyang caizheng zuo huati, difang caizheng zuo feiji* 中央财政坐滑梯,地方财政坐飞机]. In the later part of the 1950s, together with the Great Leap Forward, the entire developmental process was driven by local economies; the central government in fact did not possess the capacity to command the localities. Hence we say, for a long time since the 1950s, the local economy took precedence, it was only until the 1994 fiscal reforms that resource redistribution became half and half.

Remarkably, local autonomy was enhanced during the first decade of reforms because the-then fiscal system – which mandated a fixed percentage of total local revenue to be submitted to the central government – encouraged local governments to expand their operations but under-report profits of SOEs under their charge.[7] A Beijing-based senior planner (hereafter planner A) reveals during an interview (February 2012) these profits would often be re-categorised as expenditures in a bid to 'hide wealth within the SOEs' (*cangfuyuqi* 藏富于企). 'Between the 1980 and 1990s, planner A adds:

> the central government was so weak that there were two instances where they had to go and 'borrow money' from local governments, in each of these instances the amount was 2 to 3 billion RMB [~US$600m to US$900m in 1990 prices]. (Interview, Beijing, February 2012)

In September 1993, the-then Chinese Vice Premier, Zhu Rongji, provided a candid clarification of the fiscal position in Beijing:

> Currently central finances are confronting severe difficulties, we are in a situation where we can't go on, if we do not suitably concentrate financial revenue and enhance the financial power of the central government, the whole country will eventually be harmed and we won't be able to keep things going. (Zhu 2011: 360; author's translation)

Within a year, as detailed in Box 2.2, the CPC would unleash fiscal reforms that jumpstarted a phenomenon now widely known as 'the advancement of the state and the retreat of the people' (*guojin mintui* 国进民退).

Intertwined with this advancement is the concomitant increase in the political influence of the central government. This development corresponds with the growing theoretical debate on whether effective economic planning was a

Box 2.2   The Fiscal Origins and Contemporary Implications of Bureaucratic Entrepreneurialism

The Re-centralisation of Fiscal Control

A highly centralised fiscal system was instituted following the founding of the PRC in 1949. Provincial governments remitted their tax revenues to the central government and in turn received budgetary transfers. Direct taxation became relatively unimportant after the total nationalisation of means of production in 1958; profits from SOEs became the major contributor to state revenues. Taxation was derived from commodities sold by SOEs in its primary 'market' – the rural hinterland. This effectively meant rural residents, who already received very low wages to enable SOEs to make profits, were doubly-targeted by the state as sources of income. Urban–rural income disparities consequently became entrenched.

One common form of taxation – income tax – was instituted in 1949 but implemented in practice only in the post-Mao era. In this sense, while the central government was nominally in political control over all provinces, in practice it was very dependent on the provinces' ability to generate 'profits' through its SOEs. This created opportunities for the local governments to define what constituted 'profits', and was arguably a key reason for local protectionism and the lack of economic integration between provinces.

The situation changed following SOE reforms in the late 1980s, with the central government directly taking over the governance of some of the most important SOEs in the country (e.g. CNOOC, China Mobile, and all of the top five banks). Fiscal reform introduced some degree of decentralisation; provincial authorities had increased powers over the retention and allocation of resources. This decentralisation process augmented the provinces' capacity for planning and developing their local economies. In tandem with the devolution of developmental power to local governments (right down to the village level) to control collectively-owned 'Township and Village Enterprises' (TVEs), local governments obtained further authority in the management of finance.

However, problems began to emerge: the move left the central government short of funds, reduced its fiscal control over local authorities, and deepened uneven economic–geographical development. While uneven development was to become a driver of the CPC's 'open door policy' (i.e. integration into transnational circuits of capital flows), it swiftly became a blatant ideological contradiction.

A tax-sharing scheme (*fenshuizhi* 分税制) was introduced to increase the centre's share of total government revenue. 28 kinds of taxes were divided into three streams: one in which the central government enjoys full access, the other which flow directly to the subnational government, and one in which funds collected would be divided between the centre and the localities. Value added tax (levied on goods and services and which is the single largest source of Chinese government revenue) was shared between the central government (75%) and local governments (25%) under this scheme.

The 1994 reforms were credited with returning financial power (and hence developmental authority) from the local level to the central government, which in turn reinstated the central government's redistributive power. Yet the redistributive process remains disjointed today (see, e.g. Yep 2008; Shen et al. 2012): government agencies conduct redistributive programs independent from one another, and some of these programs contradict one another. Redistribution is also not entirely egalitarian: part of the system involves redistributing taxes in the form of rebates to provinces that pay more taxes, and these provinces are almost always the richer coastal provinces. Ironically, this system creates the perverse scenario in which poorer provinces end up paying more net taxes (when rebates are subtracted) vis-à-vis the richer provinces. For this reason, the existing redistributive system is constitutive of uneven economic–geographical development.

At the subnational level (particularly the sub-county scale), the reforms impelled local bureaucrats to be more 'entrepreneurial', as the reduction of revenues was not concomitant with the reduction of administrative responsibilities (e.g. maintenance of local roads, running of schools, etc.). This intensified the importance of 'off-budgetary revenue', a Mao-era legacy: local governments began to seek extra-fees on services that are nominally illegal; borrow indirectly from banks or the so-called 'shadow banking system' (i.e. curbside lenders that fall outside the purview of the formal regulatory system); expand the leasing of public land, etc. According to statistics compiled by Pan and Li (2011), land-financed revenue constituted 65.9% of total local fiscal revenue in 2011, a 70% increase from 2006.

As variegations in off-budgetary revenue collection and expenditure engendered new waves of uneven development and increased risks of financial instability, the CPC introduced a new measure in September 2014 to rein in the powers of prefecture- and county-level governments to launch independent developmental projects. Officially termed 'The opinion on enhancing local governments' debt management', or 'document 43' in short, this injunction 'scaled up' matters related to infrastructural development to the central government and cut off the authority of these governments to attract investors directly. Debt-financing authority was also 'scaled up' to provincial governments, which made it more difficult for these local governments raise funds. How this measure correlates to the development of more 'rational' central-planning capacities requires future research.

post-Mao phenomenon (as outlined in the preceding section). Bihong Huang and Kang Chen's (2012) study on fiscal redistribution in China reveals that whenever large scale central financial support was needed, even in the relatively prosperous coastal city-regions such as Dalian, Shanghai and Shenzhen, political connections to and/or the lobbying of the central government became necessary. Inherited from the Mao-era, this process is what the Chinese state terms 'specific purpose transfers', or *zhuanxiang caizheng zhuanyi* (专项财政转移), a redistributive mechanism that is 'typically not rule-based and thus subject to political influence' (Huang and Chen, 2012: 534).[8]

More noteworthy is the fact that these special purpose transfers constitute the largest components of fiscal redistribution from the central government to individual provinces.[9] The resultant 'anti-equalising' effects of centralised fiscal redistribution, which saw fiscal funds directed to more developed provinces, effectively generates new rounds of uneven economic–geographical development. Wen Tiejun puts the positive correlation between fiscal centralisation and the central government's political power in contemporary perspective:

Right now the central government's primary mechanism to control the local economies is fiscal transfer payments, through special purpose transfers. Naturally special purpose transfers became the central aspect of the state's financial institution, it is a major institutional gamble. To a large extent, the so-called regional disparities are caused by disparities in capital allocation [by the state]. (Wen, interview with *Nanfang Ribao*, 23 June 2013; author's translation)

The persistence of centrally influenced 'regional disparities' strongly suggests the 'cellular economy' was repurposed rather than thoroughly reconfigured after 1978 (cf. Donnithorne 1972; World Bank 1995). To be sure, market-oriented reforms introduced new processes in the cellular economic–geographical structure, namely the implementation of household-based production first experimented in the early 1960s; the gradual development of a private economy (*minying jingji* 民营经济) outside the planned system; strategic engagement with foreign capital; and the relaxation of demographic controls. Flows of goods and labour power across the country thus became less place-bound, a marked difference from the Mao-era. Despite these reforms, however, new empirical data reveals a lack of deeper inter-provincial economic integration. In 2004, the-then deputy Minister of Commerce, Huang Hai, acknowledged many local governments relied on legislative mechanisms to generate and legalise a host of discriminatory measures (*Zhongguo Qingnianbao*, 23 June 2004). In a thorough analysis of data (1978–2007) from China's National Bureau of Statistics (NBS), José Villaverde et al. (2010: 92) found inter-provincial economic disparities amidst the 'striking absence of spatial dependence, which confirms a limited economic relationship between provinces'.

Attributing this phenomenon to the restrictions on internal trade, Villaverde et al. (2010) work supports two interrelated findings by Sandra Poncet (2005), namely the fragmentation of China's domestic market along provincial lines through the 1990s and the exacerbation of this fragmentation between 1992 and 1997 (cf. Lee 1998; Young 2000). The effects of local protectionism could directly be seen in productivity disparities between export-oriented firms and firms that receive local market protection. As Rudai Yang and Canfei He (2014: 369) demonstrate, 'productive firms enjoying local protection are less likely to export' since they benefit local monopoly advantages; conversely, less productive exporting firms tend to agglomerate to enjoy 'exporting spillover effects' that allow them to directly enter international markets. In a separate study on the relationship between local protectionism and firm failure, Canfei He and Rudai Yang (2016: 80) found firms enjoying direct subsidies and loans from formal institutions are less likely to fail. The critical determinant between survival and shutdown is access to rather than the quantity of support: '[w]hether firms get subsidies and banking loans actually is more important than how much they get since both subsidies and loans represent governmental support'. These important findings underscore the constitutive effects of inherited subnational barriers to trade on the national economic–geographical configuration. By extension, it can be inferred that the 'cellular' economic–geographical structure and its accompanying protectionist tendencies that characterised Maoist China underpinned the post-Mao economic growth 'miracle' (ref. Figure 2.1).

Ironically, the new waves of local protectionism are connected to new centralising measures. In an earlier survey by the Development Research Centre of the State Council (2004), which established the widespread existence of local protectionism, local government officials justified the augmentation of local fiscal revenues through protectionism; this, as Box 2.2 shows, was a direct outcome of the central government's decision to re-centralise fiscal revenue collection in 1994. Occurring in tandem was the deepening of Deng Xiaoping's 'development is the absolute principle' (*fazhan shi yingdaoli* 发展是硬道理) mandate. Widely known in China as 'GDP-ism' (*GDP chongbai* or GDP 崇拜) and fully promoted by the Jiang Zemin-Zhu Rongji government, this approach based local officials' promotional prospects on GDP growth. Unsurprisingly, these officials relied on administrative borders to create barriers to market entry and in turn 'guarantee growth' (*bao zengzhang* 保增长).[10]

Apart from the inherited fiscal system, one other Mao-era institution – the urban–rural dual structure first introduced in Chapter 1 – underpinned the reproduction of uneven development across 'new China'. Instituted in 1958, this structure aimed to achieve three broad objectives, namely migration control, resource distribution and surveillance of targeted populations. One of the main drivers of the policy, Luo Ruiqing of the Public Security Bureau, argued that it was to 'stop the population in rural areas from blindly moving to cities'.[11] It was moot whether it had any legal basis at the time, as the draft Constitution of 1949 (*gongtong gangling* 共同纲领) and 1954 Constitution both guaranteed 'freedom of residence and movement'. Until the removal of this clause in 1975, the *hukou*

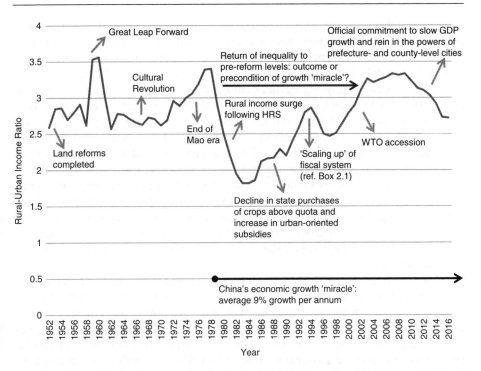

Figure 2.1   Nominal rural–urban income ratio (rural = 1), 1952–2016. Source: Based on data between 1952 and 1978 from the *Statistical Yearbook of China* (1992) and data between 1979 and 2016 from the National Bureau of Statistics (2017).

institution overtly contradicted the Constitution. For the majority of the Chinese population, however, the legality of the institution did not matter – geography just suddenly became destiny.

Of the three broad objectives of the *hukou* institution mentioned above, migration control and resource distribution play distinct functions in CPC-directed capital accumulation and are given focus in this book. The third objective remains very important today, to be sure, but it does not relate directly to the analytical focus. In addition, the objective of foregrounding the two objectives is not to evaluate the legitimacy of the *hukou* institution; that the institution had generated much social discontent for decades and had come under strong pressures for change has been well-researched (see, for instance, Chan and Buckingham 2008). Such was the severity of the social strain, the current Chinese Premier, Li Keqiang, wrote a paper two decades ago calling for the need to evaluate the 'encumbrance' of the institution (see Li 1991). Removing this 'encumbrance', however, is still not a national possibility. As the Chongqing case study in this book demonstrates, the institution has generated its own regulatory logics that benefit specific interest groups, so much so that removing it would be tantamount to removing a primary source of economic competitive advantage (ref. Chapter 7). This advantage is the surplus value generated by the previously-mentioned 'price scissors' approach to capital accumulation (*jiage jiandaocha* 价格剪刀差).

With a geographically-segregated demographic structure in place, the generation of the price-scissors effect was a straightforward process. Capital-intensive heavy industries that did not require a large labour power support base were located in the cities. To ensure these industries could receive a steady stream of funds and/or raw materials, surplus labourers were immobilised in the rural areas through large-scale collectivisation and the eventual institution of the People's Communes (ref. previous section). Under the 'unified purchase and sale' (*tonggou tongxiao* 统购统销) system, the CPC instituted a monopsony through which it could buy agricultural products at very low prices. The products were then used to sustain workers in the cities and/or used as raw materials in industrial production. This approach kept down the costs of paying industrial workers and the costs of raw materials.

At the same time, the state, acting as a monopolistic supplier, generated high profits by raising the prices of industrial products (e.g. machines, chemical fertilisers, textiles, etc.) that would then be re-sold to the communes. Through accepting state-suppressed prices for their agricultural products and paying higher prices for goods produced by urban-based industries, peasants were effectively subsidising national-scale industrialisation in China. According to observations by Li Changping (2008: n.p.; author's translation), a renowned specialist on agricultural issues in China, the state-driven industrialisation project effectively meant re-instituting a system of involutionary expropriation:

> The acceleration of national industrialisation quickly led to a dependence on the institutions of absolute centralisation and high-level planning that expropriated peasants – in an involutionary fashion [*neijuanhua* 内卷化], they triggered primitive accumulation and enabled state capitalism to complete its industrialisation. Peasants were forced to lose their individual autonomy during the process of national industrialisation, they became tools of nationalised industrialisation.

And, as Li (2008: n.p.) adds, these 'tools' served the nationalised industrialisation project well:

> In not too long a period, the 'price scissors' became a major achievement of industrialisation, China developed a complete and autonomous industrial system, agricultural system, national defense system, educational system, cultural system etc., it even recovered its place in the UN. The nature of and right to national autonomy was fully realised, and the history of colonisation underwent a fundamental change.

To Li Changping, this 'primitive' capital accumulation process was predicated on and reproduced a form of internal colonisation. Driven by the two national-scale spatial projects of People's Communes and the *hukou* institution, the reproduction of colonisation within Chinese state spatiality precluded peasants from enjoying the gains that come with individual householding:

> This manner of sacrificing peasants' autonomy in exchange for national autonomy should, in Mao's initial state-building thoughts, have ended in the 1960s. It should have evolved from 'peasant supporting industrialisation' (*yinong bugong* 以农补工) to 'industrialisation supporting peasants' (*yigong bunong* 以工补农). Yet up to the time of Mao's passing, the industrialisation strategy of expropriating peasants had not changed. Industrialisation ultimately met the passive resistance of the peasants – they expended labour but not vigour, lacked creativity and efficiency, and ultimately led to the ineffectiveness of the People's Communes institution. The eventual outcome was the inability of industrial goods to be exchanged for goods of good quality from the countryside, leading to a crisis of national industrialisation…The failure of industrialisation between 1949 and 1978 was practically a failure of a developmental trajectory based on the internal colonisation of peasants. (Li 2008: n.p.)

Yet, in an ironic twist, Deng Xiaoping initially addressed this crisis on the basis of this dual structure. This was achieved through two parallel developmental paths, namely (i) the spatially selective 'opening up' to global circulatory capital in the southeastern SEZs and (ii) the reconfiguration of economic production in the rural hinterland, within which more than 80% of the national population resided at the time, from the commune level to the level of the individual household (in what remains known as the 'Household Responsibility System' [HRS], or *jiating lianchan zerenzhi* 家庭联产责任制). Yet the unrest that unfolded during the first decade of reforms demonstrates both paths could not run independently. This led to a two-tiered transformation in the 1990s: the *convergence* of these paths and a concomitant prioritisation of the urban (cf. Andreas 2010; Wallace 2014).

Finance and assets began to be concentrated by select private owners with the expanded privatisation of rural industries, in particular to insiders previously working in the rural industries (Hinton 1990; Rozelle et al. 2000; Li and Rozelle 2004). This corresponding search for 'economic efficiency' through 'crowding out' produced rural class polarisation and increasing inequality, with a growing number of rural workers either laid off or forced to accept wages significantly lower than those offered in the coastal city-regions (So 2003; Liu 2006; Webber 2012). Higher wages in the cities took on new significance with the rollback of rural social services and the decline in rural employment in the same period. As a senior planner in Beijing puts it, the contraction of these capacities was correlated to the dismantling of the People's Communes in and after 1984:

> To take as an example, under the commune institution, if an instruction came in from above to build roads, then the commune cadres must think of ways to get the funds through economic production before using the funds to build the roads. Under the current institution, similarly instructions will come from above to build

roads or provide other facilities, but the village governments have no money. Because they are not in charge of economic production, they will go around assessing people in the villages for funds, this easily creates situations where funds are extracted illegally, without regard for the necessity and ability of the villagers to pay. Nowadays they even have to borrow from banks or non-authorised agencies, but these [loans] must be paid back. From whom do they get the money? The peasants. (Interview with senior planner B, February 2012; author's translation)

An important observation of this planner is the apparent re-emergence of 'bureaucratic capitalism'[12], a trait the Mao administration viewed as a primary target of the pre-1949 revolution:

> The village governments have become a bureaucratic institution (*guanliao tizhi* 官僚体制), they rear a bunch of people who have nothing to do with economic production, hence their activities directly burden the peasants. If you look around, what we have today is really a form of bureaucratic capitalism (*guanliao zibenzhuyi* 官僚资本主义). (Interview with senior planner B, February 2012, Beijing; author's translation)

This two-pronged development – the reduction of agriculture-related jobs and the re-emergence of 'bureaucratic capitalism' in the rural hinterland – arguably made possible what Barry Naughton (1995) terms 'growing out of the plan'. Corresponding with Minxin Pei's (2006) observation of 'selective withdrawal', the 'growing out' process was characterised by a 'dual track' approach to production initiated by the former Premier, Zhao Ziyang, in 1984. This approach replaced the Soviet-styled 'planned system' with the defining feature of post-Mao socioeconomic reforms – marketisation. While some parts of industrial production remained planned by the central state and anchored the national economy, certain aspects of previously monopolised industrial sectors were opened up to private buyers and sellers. Surplus goods could be sold at privately determined prices and producers in these sectors were able to compete directly with the growing non-state economy. The overarching objective, to Barry Naughton (1995: 200), was to 'expand market forces by limiting the scope of planning, fostering entry, and improving incentives and autonomy for state-run enterprises to operate on the market'. 'Dual track' production culminated in an official acknowledgement in 1993 that all means of production should be geared towards the development of a 'socialist market economy'. This economically nationalistic project is now officially termed 'the China Dream'.

The gradualism of 'growing out' distinguished the Chinese transition to market-like rule from the 'big bang' policies of the former Soviet Union and Eastern Europe in one major way: there was no corresponding political instability at the national scale. However, one crucial precondition requires emphasis: the 'planned system' Naughton (1995) referred to was effectively an urban-based system. It was a system institutionally, structurally and socially separate from the rural hinterland. This distinction means two-track pricing would be hard to implement in the former Soviet bloc because the state industrial sector in this bloc played a dominant role and the majority of the population was employed by state(-linked) institutions. In contrast, only 18% of China's population were based in urban-based industrial *danweis* (place of employment) in 1978. Contrasting the pressures for total socioeconomic restructuring in the former Soviet bloc, the CPC did not have to induce unemployment in order to facilitate market-based reforms. Zhao's 'dual track' approach was therefore capable of driving non-state economic growth because of Mao's apartheid-like policy to segregate the rural population (and its associated economic activities) from the urban. With state monopoly relaxed in the industrial sectors, the platform was established for the rural-based economic actors to engage in direct competition with urban-based SOEs. TVEs gained market entry and stimulated competition (Naughton 1994). At the same time, the approach also facilitated the launch of the biggest non-state market of all – the market for surplus rural labour power.

If there was a primary corollary of post-Mao marketisation, it was the seamless production of a 'floating' labour market that powered the exponential GDP growth in China since the early 2000s (cf. Chan and Buckingham 2008). Within this market, peasant migrant workers could be priced solely as social labour power. Since the early 1990s, this has taken place in segmented intra-urban labour markets comprising what Cindy Fan (2002: 103) terms 'the elite, the natives and the outsiders'. Without corresponding socio-spatial protection, these 'outsiders' became less rooted to their new urban destinations; yet without corresponding growth in employment opportunities in their rural 'hometowns', they could only continue to 'float' between city-regions to seek employment (Solinger 1999; Chan 2010a, b). Peng Xizhe, a leading public policy scholar from Shanghai's Fudan University, puts the situation in perspective:

> The reason why rural workers can engage in economic activities at much lower wages in the cities is due to this two-tiered social status. If there is no hukou institution in China, rural workers will hope to have the same lifestyles, the same wages, and the same working conditions as those living in cities. Yet because one continues to feel s/he is a peasant [in the cities], that the 'home' is in the countryside, the feeling of being in the city would be for the purpose of earning some income. So long as wages are acceptable [in relation to those in the countryside], s/he will be willing to work. This phenomenon is an important precondition of the existence of low-cost labour in China. (Peng, interview with *Xinhua*, 13 August 2009); author's translation

Two distinct functions constitute this important precondition. First, it eases the financial burden on municipal governments to provide social welfare. This consequently allows these governments to concentrate on attracting and developing factors of production necessary to embed capital. Second, it ensures the labour supply is geographically elastic to shifts in effective demand for labour power (which to a large extent is derived demand, due to the export-orientation of many labour-intensive industries along the coastal seaboard after the 2001 World Trade Organisation (WTO) accession). The second economic function is particularly critical in times of crisis because, in reaction to crisis induced unemployment, the *hukou* institution makes it administratively possible to mandate migrants to return to their place of household registration. He Xuefeng, director of the China Rural Governance Research Centre in Huazhong University of Science and Technology, crisply describes the contemporary function of the dual structure:

> At the current stage of development in China, and viewed from China's position in the global division of labour, there will be a large underclass that has low pay and very uncertain employment conditions. Because of the existing structure, this underclass could still move freely between the urban and rural areas. (He, interview with *Nanfeng Chuang*, 14 January 2014); author's translation

The socioeconomic logic of retaining rural 'territorial absorbers' is highlighted by the Guangdong provincial government's swift response to the 2008 global financial crisis. As Chapter 4 will elaborate, a 'double relocation' industrial policy (*shuang zhuanyi* 双转移), also known as 'emptying the cage to change the birds' (*tenglong huanniao* 腾笼换鸟), was instituted to relocate unwanted industries and the labour power they employ (the 'birds') away from the Pearl River Delta (the 'cage'). While it was expected that this restructuring would result in a temporary dip in economic performance, the official employment figures in Guangdong province were positive even after the crisis struck in 2008. Nationally, the published unemployment rate was under 10%. This rosy picture was complicated after Wen Jiabao, the-then Premier, acknowledged in a meeting with foreign delegates in 2010 that 200 million people were unemployed (*China Daily*, 23 March 2010). Relative to the total population (~1.38 billion), this translates to a 14.5% unemployment rate; the rate becomes 20% when measured against the total working population (~1 billion).

In relation to official accounts from Guangdong, Wen's speech strongly suggests unemployment that should be concentrated within Guangdong, the most attractive destination for migrant workers across the country when the crisis struck in 2008, was successfully 'relocated' to other provinces. This 'exportation' not only explains why 200 million people (a substantial number who would have been working in the Pearl River Delta) were unemployed across China in the aftermath of the crisis while the official employment growth rates remained positive in Guangdong. Viewed at the national scale, it foregrounds the capacity of the 1958 urban–rural dual structure to subordinate the 'floating population' to the demands and crisis tendencies of transnational circulatory capital (of which the CPC is now an integral part), just as their subservience during and after the crumbling industrialisation efforts strongly surprised Mao in the 1960s that he exclaimed: '*Hao ah!* [Marvellous!] You whistle and 200 million people come, you wave and they go, without the CPC in power, which party can achieve this?!' (*People's Daily*, 14 October 2013; author's translation)

It becomes evident, then, that the 'growing out of a plan' or 'selective withdrawal' process in post-Mao China was not just about the relaxation of state monopolies in the hitherto urban-based 'planned system'; the 1958 urban–rural dual structure was a vital enabler. More importantly, it continues to underpin contemporary industrial policies. This re-affirms Pei's (2006: 26) observation that the gradualist approach worked because 'it has allowed Chinese leaders fully to exploit the structural advantages provided by favorable initial conditions'. striking feature is its continued reliance on the Maoist approach to 'use the rural to support [urban-based] industrialisation' (*yinong bugong* 以农补工). As He Xuefeng (2010), Kam Wing Chan (2010a) and Fulong Wu (2015) have shown, growth-oriented planning occurred in tandem with the strategic repurposing of the dual structure.

Taken together, the literature on uneven development in post-1949 'new China' illustrates perpetual CPC rule as a contingency – a socio-spatial contingency. At the inception of reforms in the 1980s, the Deng Xiaoping regime had to work with or around the previously mentioned regulatory logics of the Mao era. Breaking off these legacies too suddenly could unsettle the nascent coherence of a political structure painstakingly built up by the Mao administration. For this reason, experimental policies were introduced through the HRS, SEZs, and more recently the 'nationally strategic new areas'. Identifying transformative change within this path-dependent context thereby requires a robust re-evaluation of uneven development in post-1949 China.

To begin, it would be helpful to move beyond measuring whether inter-regional inequality is inherently negative or unethical vis-à-vis the CPC's self-proclaimed quest for socialism. As this section has shown, institutional change within the sprawling party-state apparatus is not constituted by a one-track movement from spatial egalitarianism in the Mao era towards greater fragmentation in the 'transitional' present. Rather, as presented in Table 2.1, the evolutionary process is characterised by the interaction of multiple regulatory layers, each with their distinct geographical expressions, with inherited institutions. Chinese politico-economic development is thus more accurately a complex palimpsest: *some Mao-era logics of socioeconomic regulation (like the People's Communes and the 'permanent revolution' campaign) may have gone, but others have been re-purposed through new institutions to facilitate market-like rule (see far right column)*.

Conclusion

Is China an exceptional case because its Leninist system has been able to sustain change along the lines of a mixed economy? Or is Chinese exceptionalism something of an illusion which obscures an underlying conformity with the fundamental logic of a Soviet-style system – the fusion of politics and economics? It is quite possible to give affirmative answers to both questions. – Steven Goldstein (1995: 1110)

Table 2.1 Post-1949 Chinese politico-economic evolution as a complex palimpsest.

Regulatory layers	Spatio-temporal characteristics	Regulatory objectives	Key capital source	Repurposed institutions from the Mao era
Recalibrating uneven development (Mid-2000s to present)	• Takes place through simultaneous policy experimentation in 17 institutionally-differentiated 'nationally strategic new areas' at time of writing (September 2017) • Became apparent in the mid-2000s, as socio-spatial disparities widened between provinces and within city-regions • Labour-intensive industries encouraged to move to interior city-regions; coastal city-regions to 'transform and upgrade' their positions in global production networks (*zhuanxing shengji* 转型升级)	• Retain transnational circulatory capital within Chinese state spatiality • Deepen integration with the global financial system • Capture cost advantages of inland city-regions, particularly access to low-cost labour and land • Unleash domestic demand through rural–urban migration	• State-led fixed capital investments (debt-financed) • Expanded domestic credit creation targeted at consumers • Foreign direct investments (FDIs)	• Hierarchical regulation of state spatiality through the Lenin-styled *nomenklatura* and informal patron-client personnel structure • Quasi-cellular economic organisation remains, characterised by local economic protectionism and inter-provincial trade barriers • *Hukou* institution determines control of labour power flows • Centralised financial capital controls, including extensive state participation in the domestic financial system and commitment to intervene, whenever 'necessary', in financial markets 'without hesitation' • State- and collective landownership allows the central state apparatus total control over the use- and exchange-value of land transfers at the national scale • Centrally-controlled SOEs dominate key economic sectors • Industrial policy remains a key regulatory tool

| Urbanisation of capital and labour power (1980s to present) | • Originated from the 1980 launch of coastal Special Economic Zones
• Distinct urban developmental bias became clear in the early 1990s, as deepening rural privatisation and decollectivisation of socioeconomic life created socioeconomic pressures to absorb rural surplus labour
• Further entrenched urban-scale 'local state corporatism' or bureaucratic capitalism (He and Wu 2009; Wu 2015)
• Central-local interaction began to shift from provincial to major centres of real estate development in leading city-regions (Tsing 2010) | • Increase overall economic productivity
• Embed global production networks
• Enhance competitiveness of SOEs | • FDIs
• Tax credits
• State finance
• Private equity financing (through domestic and foreign stock exchanges) |

(Continued)

Table 2.1 (Continued)

Regulatory layers	Spatio-temporal characteristics	Regulatory objectives	Key capital source	Repurposed institutions from the Mao era
Rural 'roll back' (1980s to early 2000s)	• The precondition of China's economic 'miracle'; occurred parallel to SEZ formation and 1984 'opening up' of select coastal cities • 'Household Responsibility System' (HRS) launched in 1980 to select provinces (e.g. Anhui, Sichuan and Guangdong) but established nationwide by 1984 • Decollectivisation in 1984 triggered the rollback in social welfare provision in the vast rural hinterland, where 80% of the national population resided in 1980 • Rural privatisation continues to deepen through the 2000s, with large corporations increasingly 'crowding out' small producers (Webber 2012)	• Increase agricultural output • Increase individual incentive to farm • Reduce state burden on welfare provision	• State investments in agricultural infrastructure • Domestic capital (private and SOEs/collectives)	

National politico-economic integration (1949–1978)	• Economic space treated as a 'chessboard' by Chinese policymakers since the Mao-era (Zhao 2009) • People's Communes formed to universalise the formation of production units in the rural hinterland; heavy industrial units (*danweis*) concentrated in cities, supported by rural economic production • Inter-unit interaction limited and controlled by the central government through mobility restrictions and trade barriers introduced on the premise of 'self-sufficiency' • National economy became 'cellular', held together by a centrally-determined spatial hierarchy (Donnithorne 1972; Ma 2005)	• Reproduce a Stalin-styled socialist economy capable of 'catching up with the UK and US' • Reinforce the political power of the CPC over the newly-established national state space	• Aid and favourable trade policies from the Soviet Union (up until 1960) • Extraction of surplus value through 'price scissors' mode of accumulation

In carrying out the construction of socialism by our party, leaders, and people, there are two periods, [one that is] before reform and liberalisation and [one] after that... Although there were major differences in the ideological direction, orientation, and policies in the implementation of socialist construction during these two historical periods, they cannot be disconnected. Furthermore, they are not oppositional. One cannot use the historical period following reform and liberalisation to negate the historical period prior to reform and liberalisation, and vice versa. – Xi Jinping, President of China. (*People's Daily*, 6 January 2013; author's translation)

When Mao Zedong told Poul Hartling in 1974 that efforts at building a socialist state used and reproduced the methods of the 'old society', he was arguably aware of the limits to driving capital accumulation on the basis of instituted uneven development and geo-economic insulation (ref. Section 2.1). Yet, in selecting Hua Guofeng (rather than Deng Xiaoping) as his official successor, Mao highlighted the importance of retaining Stalinistic logics of socioeconomic regulation the CPC so painstakingly cultivated in the 1950s and 1960s.[13] While it might appear at first that Deng's success at bypassing Hua to launch market-oriented reforms in 1978 marked a clear separation from Mao's reign, subsequent research on the interaction between Mao-era regulatory logics and successive waves of post-Mao reforms presents a more nuanced picture of continuity and change. This chapter has critically evaluated this interaction from a geographical perspective.

Questioning the notion that socioeconomic reforms in the post-Mao era represent a linear-sequential devolution that transformed the Chinese political economy, this chapter demonstrates how the reproduction of national-level regulation was not compromised by one-directional movements towards decentralisation and uneven development. Contrasting the drastic changes in the former Soviet bloc, Deng and his successors repurposed central political institutions *through* decentralised governance, territorially targeted policy experimentation and the reproduction of regulatory uneven development. Rather than constitute a one-track shift from centralised state-socialism to a decentralised market regime, changes that took place produced a quasi-exceptional process in which Leninist socialism *co-exists* with marketisation to produce new spatial logics of socioeconomic regulation (cf. Howell 2006; Wei 2007; Li and Wu 2012; Lim 2014). There is no eclipse of one national 'system' (or paradigm) by another.

Indeed, new reform initiatives and institutional continuities are co-constitutive. As the current Chinese President, Xi Jinping, acknowledges, the political economic history of 'new China' cannot be characterised by two distinct and oppositional 30-year periods. While there were clear differences in the ways economic production and demographic flows were regulated by post-Mao regimes, the primary regulatory objectives of these regimes arguably did not differ from Mao's (ref. Sections 2.2 and 2.4). These regimes all sought to deepen the foundational status of the CPC as the government of China against continuous threats to regime stability (e.g. the Korean war, fears of Kuomintang resurgence, the Tiananmen protests, ethnic tensions in Xinjiang and Tibet, etc.) and all pursued economic nationalism through state-led capital accumulation. What has changed were the methods – or, more specifically, the *spatial strategies* – to achieve these objectives. Mao chose the 'cellular' organisation of the People's Communes, rural industrialisation and geo-economic insulation, even to the extent of shutting out Soviet influence during the 1960s; Deng and his successors sought to engage and ultimately integrate with the global system of capitalism through policy experimentation in targeted territories, the most recent series being the 'nationally strategic new areas' (Bramall 2007; Day 2013; Lim 2014).

While these approaches generated contradictions and led to new reforms, many inherited institutions were never fully abandoned (ref. Table 2.1, far right column). For Mao, the aftermath of the Great Leap Forward could be superseded through successive rounds of political campaigns, the resolute suppression of household-based production and the retention of key regulatory logics of Stalin's regime, primarily the harsh treatment of dissenters, the collectivisation of production and the use of the 'price scissors' approach to capital accumulation (ref. Section 2.4). In response, Deng and his successors instituted market-like rule while reinforcing at the same time the central government's politico-economic control. This was primarily characterised by the introduction of individual initiatives in economic production in the rural hinterland (which has since been extended to the cities); the launch of successive rounds of geographically targeted experimentation to embed transnational circulatory capital; the simultaneous reconfiguration of key Mao-era institutions such as collective/state landownership, the urban–rural dual structure and a financial system dominated by state-linked institutions; and, last but not least, the readiness to intervene in market functions 'without hesitation', as exemplified by its decisive attempt to dictate stock market trading in the summer of 2015 (see far right column in Table 2.1). It is apparent from these developments that the enlargement of capital, which encompasses both the state and non-state sectors, has been entwined with its subsumption to party goals in the post-Mao era. This reinforces Jun Zhang's (2013: 1614) observation that 'the political power of capital in China remains fundamentally embedded in, and interlaced with, the sprawling institutional machinery of the Leninist party-state and the political capacities of the CCP'.

It would be more apt, then, to conceptualise politico-economic evolution in post-1949 'new China' as a *cumulative* process through which place-specific experimental policies are layered on inherited logics of socioeconomic regulation (ref. Figure 2.1). Emerging from this regular re-layering is a multi-scalar state spatiality that is connected to the global economy in highly uneven ways. In turn, this unevenness determines whether the CPC can continue to govern the Chinese political economy as a unitary and hierarchical political system. The old 'planned system' may have given way to market-like governance, but the logics of the market have become the *baseline* from which new spatial strategies are developed by the party-state apparatus. While 'withdrawal' has taken place in 'select' domains, as Pei (2006) rightly points out, there is reinforcement of state power in others (see Zhang [2013]; Zhu [2013]; and Wu [2015]). The rollback of political capacities may be apparent when it is viewed in relation to the inherited and seemingly static institutional template (e.g. HRS vs. Maoist collective production), but it is never a zero-sum process. Change is instead constituted by the complex relations between institutional path-dependency, policy experimentation and the proactive reconfiguration of regulatory scales by subnational political actors in the name of the 'national strategy'. Indeed, if Wen Tiejun and other Chinese 'New Left' scholars are correct in claiming that central socioeconomic planning is only blossoming in the current conjuncture, the crucial empirical question would be whether and how this planning could work *through* an overarching process the CPC established since Deng's reforms, namely the feedback loop between territorially targeted policy reforms, global economic–geographical competitiveness and central(ised) political power. The next chapter will develop a conceptual framework to explore this process.

Endnotes

1 That these three commodities, deemed 'fictitious' by Karl Marx, were placed under public ownership is in itself symbolic.
2 This re-connection of SOEs' management strategies with political goals corresponds to the CPC's recent rejection of plans to depoliticise state-owned enterprises by governing them through financial holding companies (*Financial Times*, 20 July 2017).
3 With the implementation of the 12th Five Year Plan in 2012, the Chinese central government has regularly committed to reduce its GDP growth target to a 'new normal' of around 8%. This figure was revised to 7% in 2015 as the CPC, in Premier Li Keqiang's terms, sought to fight "systemic, institutional and structural problems" (*The Guardian*, 8 March 2015). A major policy to address this problem was launched in September 2014 and was targeted directly at reducing the powers of prefecture- and county-level cities (see elaboration in Box 2.1).
4 For the full text of this injunction in Mandarin, see *People's Daily* (1 August 2003). Current English text is translated by the author.
5 The program was to accelerate the pace of industrialisation by making use of cheap and abundant labour in the countryside while sidestepping the importation of machinery.
6 While the primary – and well-documented – expressions of economic nationalism by the CPC during the Mao era was its aim to 'catch up with the UK and surpass the US' (*zhuiying ganmei* 追英赶美), a more thorough historical exploration reveals its ultimate goal could be to match, if not surpass, the Soviet Union. As Mao puts it tellingly in December 1957: 'We can only copy the Soviet Union in the fundamental sense, even so I feel dissatisfied, I feel uncomfortable…The Soviets produced more than 4 million tonnes of iron in 1921, this increased to 18 million tonnes in 1940. Both the Soviet Union and China are socialist countries, could we not accelerate and expand production, could we not use a method that is bigger, faster, better and more economical to build socialism?' (cited in Deng 1998: 44 and 715; author's translation).

7   The situation was exacerbated by a new initiative known as '*baoying bu baokui*' (包赢不包亏) that encouraged the retention of profits but not responsibility when losses were incurred. This institution was subsequently reformed as SOEs were mandated to function more like private corporate entities.

8   'Special purpose transfers', Huang and Chen (2012: 538) explain, are 'commonly used by the central government to provide incentives for local governments to undertake specific policies, programs, or activities favoured by the central government'. From 2007 data, for instance, it could be seen that while many poor provinces like Tibet, Guangxi and Jilin have benefited from these transfers, a high amount also goes to Shanghai, one of the richest province-level areas in China. Through enhancing the development of Shanghai, coastal-interior unevenness was reproduced, if not reinforced. See also Sheng (2010).

9   The cities of Beijing, Chongqing, Shanghai and Tianjin are classified as province-level administrative units directly governed by the central government (*zhixiashi* 直辖市). Under this classification, these cities are treated at the same level as provinces like Guangdong and Fujian.

10   A rich vein of research has been conducted on the transfers of administrative personnel around provinces under a *nomenklatura* personnel system; see, for instance, Naughton and Yang (2004), Chien and Gordon (2008), Xu (2011).

11   Prior to 1958, the CPC began calling rural migrants into the cities 'blind flows' (*mangliu*); the implication is the rural migrants had no consciousness of 'reality' and were just 'blindly' going along with the 'flow'. It is intriguing how, in the contemporary context where almost a seventh of the population constitutes 'floating population' (i.e. migrants in locations not of their *hukou* registration), seemingly incessant flows of rural populations into cities are no longer termed as 'blind flows'. By implication, then, the contemporary flows are calculated decisions, rationally responding to price signals in the cities. Yet one thing remains unchanged: whether 'blind' or rational, the inflows of rural migrants would not be entitled to social benefits.

12   Given the constraints of space, it would not be possible to elaborate on the concept of 'bureaucratic capitalism' and its empirical manifestations in post-1949 China. By definition, this refers to economic rent-seeking from political actors through the (ab)use of political powers. Mao Zedong made it clear in March 1948 that 'bureaucratic capitalism' was to be 'overturned' together with feudalism and imperialism in the Chinese revolution (Mao 1991: 1182).

13   Hua Guofeng would immediately declare a 'Two Whatevers' (*liangge fanshi* 两个凡是) approach to managing Mao's legacies: (i) We will resolutely uphold whatever policy decisions Chairman Mao made; and (ii) We will unswervingly follow whatever instructions Chairman Mao gave. This approach indicates Hua would have extended the quasi- Stalinistic regulatory logics across China if he had remained in power. Hua was, however, out-manoeuvred by Deng and his strong support network in 1978.

Part II
Conceptual Parameters

Chapter Three
State Rescaling, Policy Experimentation and Path-dependency in post-Mao China
A Dynamic Analytical Framework

It is now broadly accepted that processes of contemporary state restructuring are deeply entangled with transformations in scalar relations. What is at stake here is far more than rescaling for rescaling's sake, because these forms of scalar restructuring are both a medium and an outcome of changes in the means and ends of state action. Hence the need to move beyond 'thick descriptions' of state restructuring, policy reforms and new forms of governance to ask what it is that the state is actually doing – why, where and with what political, social and economic implications. The careful mapping of emergent state forms can, and should, be an important part of this process.

– Jamie Peck (2003: 222)

Introduction

This book frames the emergence of 'nationally strategic new areas' across China as dynamic entwinements between state rescaling, place-specific policy experimentation and institutional path dependency. Initial conceptualisations of state rescaling were not drawn from developments in China, to be sure. Rather, scholars first noted how the transition from a nationally-configured, Fordist–Keynesian developmental approach in western Europe and North America was underpinned by a corresponding reconfiguration of regulatory relations between national, subnational and supranational governments (Peck 2002, 2003; Brenner 2004; Jessop and Sum 2006). Emerging from this reconfiguration are 'post-national' states

On Shifting Foundations: State Rescaling, Policy Experimentation and Economic Restructuring in Post-1949 China, First Edition. Kean Fan Lim.
Published 2019 by John Wiley & Sons Ltd.

that simultaneously (i) embrace transnational economic integration and (ii) establish city-regions as 'new state spaces' of/for capital accumulation. Because a similar reconfiguration was occurring in previously-insulated China following the re-introduction of market-like rule in 1978, the state rescaling framework has more recently been adapted by researchers seeking to evaluate and explain the reproduction of Chinese statehood.

Featuring prominently in this emerging research stream is the constitutive and at times conflictual role of city-regionalism. As Yi Li and Fulong Wu (2012, 2018) show, competing interests encumbered the central government's unilateral implementation of the Yangtze River Delta regional plan (see also Wu and Zhang 2007; Wang et al. 2016). Locally differentiated definitions of citizenship, inter-city collaboration and growth have similarly generated tensions in the urbanisation of the Pearl River Delta (Smart and Lin 2007; Yang and Li 2013; Li et al. 2014, 2015; Sun and Chan 2017). Yet state rescaling was also characterised by centrally-driven 'upscaling', as shown in the re-designation of an entire province, Yunnan, to enhance engagements with the Mekong sub-region in Southeast Asia (Su 2012a, b; cf. Tubilewicz and Jayasuriya 2015). Underpinning these studies is a common, if not always explicit, assumption: the threefold engagement with marketisation, geo-economic liberalisation and economic globalisation has necessitated the recalibration of national-level regulation in China. By extension, previously subordinate regulatory scales – namely the collective, municipal, county, prefectural, provincial, cross-provincial, and transnational – have become functionally and strategically more important.

This chapter evaluates Chinese state rescaling on the basis of this assumption. While concurring with the prevailing literature that state rescaling offers a useful platform to examine the relationship between subnational transformation and state restructuring in China, the chapter is also cognisant of its conceptual limitations as identified in existing studies. To Li and Wu (2012: 58), the rescaling framework 'does not offer an effective tool for examining the causal relationships and dynamic processes of the changing statehood'. Surveying the broader state rescaling literature, Zhigang Li et al. (2014: 131) identify an additional 'tendency to apply a hegemonic interpretation of state space theory, at the expense of knowledge of place-specific practice'; correspondingly, there is 'a lack of detailed case studies revealing the key political processes and relationships that reflect historical contingencies and path dependencies under transition'. For this reason and considering the Chinese institutional context, Yi Sun and Roger Chan (2017: 3285) argue that the 'genesis' of each city-regional new state space 'deserves a meticulous examination'.

Building on these critical insights, this chapter shows how studies of post-1978 state regulatory reconfiguration in China overlook a fundamental aspect of the original state rescaling framework – the crisis of and subsequent tensions with the preceding regulatory institutions. Indeed, state rescaling was originally portrayed as a strategic response to the limits of Fordism as an accumulation regime; as a shift from a nationally-oriented, geographically egalitarian mode of surplus redistribution known as 'spatial Keynesianism' towards more flexible, 'post-national' accumulation in city-regions (Brenner 2004, 2009; Jessop 2016). Fresh debates subsequently emerged as research demonstrates city-regionalism was not accompanied by clean breaks with the legacies of Fordist-Keynesianism (ref. Peck 2003; special journal issues edited by Lobao et al. 2009 and MacLeavy and Harrison 2010). The dynamic relationships that produced and constituted the complex palimpsest as presented in Chapter 2 are largely unaddressed in current conceptualisations of Chinese state rescaling, however. This oversight is problematic given the significant influence of Mao-era institutions in spite and in some instances even *because* of the shift to market-like rule. A more incisive analytical framework that foregrounds and examine these relationships would therefore advance existing knowledge of Chinese state rescaling.

To this end, the chapter moves beyond western-based studies by situating Chinese regulatory reconfigurations, most recently expressed through the designation of 'nationally strategic new areas', vis-à-vis two cognate drivers of political-economic evolution – policy experimentation and path dependency. Aligning with Peck's (2003: 222, as quoted at the beginning of the chapter) caveat not to focus on 'rescaling for rescaling's sake', this framework encourages a 'careful mapping of emergent state forms' by asking why and how regulatory changes (primarily effected through experimental policies in selected city-regions) *interact* with inherited pathways (the retention, if not reinforcement, of national-level institutions). Rather than periodise regulatory reconfiguration as static shifts 'upwards' or 'downwards' from the national scale, the framework sets up the analysis of 'nationally strategic new areas' as a *politicised process shaping institutional continuity and change at different spatial scales*.

The discussion is divided in three parts. The following section critically reviews the conceptual origins of state rescaling and delineates four reasons for reconfiguring the existing framework in this book. This sets up the presentation of an integrated analytical framework and research agenda that places state rescaling in dynamic engagement with place-specific policy experimentation and path dependency. The chapter closes by highlighting the implications of the new framework for the analysis in the subsequent chapters of this book.

State Rescaling and Socioeconomic Reforms in Post-Mao China: A Critical Overview

The concept of state rescaling was originally developed to explain changing spatial divisions of regulation in western Europe following the Fordist–Keynesian regulatory crisis in the mid-1970s. Underpinning this crisis was a move *away* from a political commitment to full employment and (relative) regional equality, or what is widely known as spatial Keynesianism. In its place were rolling urbanising strategies that qualitatively modified the nationally-oriented regime of capital accumulation. While these strategies generated new growth opportunities, they similarly contained fresh crisis tendencies and therefore generated *recurring* rounds of rescaling. As Brenner (2009: 127) reflects on *New State Spaces* (2004), a major book-length reference point for the state rescaling approach, 'urbanisation processes would engender contextually specific forms of sociospatial dislocation and crisis formation, as well as corresponding strategies of political intervention designed to confront the latter'. Brenner (2009: 128) then clarifies the connection between this recurrence and regulators' inability to contain crises:

> since the 1990s, new forms of state rescaling have emerged largely in response to the crisis tendencies engendered through the first wave of urban locational policy. This has led to the construction of new scales of state intervention (neighbourhoods, metropolitan regions and transnational interurban networks), to the crystallisation of additional crisis tendencies and dislocations and, subsequently, to a further intensification and acceleration of rescaling processes. Processes of state rescaling therefore appear to be animated through regulatory failure.

Primary aspects of 'regulatory failure' in western Europe were encapsulated within the dismantlement of what Jessop (1993) terms the 'Keynesian welfare state'. These aspects were the crisis of the welfarist state system; the 'internationalisation' of previously-Fordist corporations; and the 'hollowing out' of the state (cf. Rhodes 1997; Peck 2001). For Jessop and Sum (2006: 271, 281), the decline of the national scale as the 'taken-for-granted object of economic management' across western Europe, the East Asian 'trading nations' and important-substituting Latin America marked the emergence of a 'relativisation of scale' in socioeconomic regulation, namely 'the absence of a dominant nodal point in managing interscalar relations'.

This 'absence' consequently led to the emergence of what Peck (2002) terms 'inter-scalar rule regimes'. While states remain key actors under this new arrangement, Peck (2002: 340) observes they are now engaged 'in more active processes of scale management and coordination at the local and international levels'. This management and coordination process is multi-directional; agendas espoused by subnational and supranational regimes influence national regulation in strategic and dynamic ways (Jones 2001; Harrison 2010; Jonas 2013). The predominant objective and outcome of western European state rescaling was and remains the negotiation between actors in these rule regimes to concentrate capital in and across selected city-regions (Ward and Jonas 2004; Cox 2009, 2010; Harrison 2012). This raises the question whether the structural coherence underpinning Fordist Keynesianism – and, indeed, of nation-states *in general* – has been completely transformed.

Structural coherence extends David Harvey's (1985) concept of 'structured coherence'. By Harvey's formulation, capital seeks to accelerate the time it takes labour to convert commodities into more commodities and profits. This process is premised on geographical transformations in and through nation-states (hence transnational circulatory capital hereafter). The places must be (re)produced to coordinate capital circulation through the concentration of infrastructure, transport connections, housing for labour power, factories and consumer markets. Harvey (1996) subsequently terms such places 'permanances', namely the provisional stabilisation of investments and processes in place. Once stabilisation is achieved, a particular place would enhance the ability of capital to extract value from labour power and accelerate the exchange of goods from plants to final markets. The intrinsic contradiction of structured coherence is it contains conditions that could undermine the 'permanences' that constitute this coherence. Firms and governments elsewhere could try to outperform the dominant locations of accumulation, or firms in the 'permanent' territories could seek to leave for places that could potentially generate more surpluses. Any sort of coherence that emerges is therefore intrinsically unstable.

Jessop develops this point to evaluate transformative change between regulatory regimes: change would not have undermined structural coherence if the dominant scale of accumulation – which, in the case of North America and western Europe in the post-WWII period, was the 'nation-state' – has remained stable *despite* disruptions (Jessop 2001, 2002; Peck 2001; Brenner 2004). The crucial variable rests in the ability of changes to disrupt the structural coherence of this dominant scale. Policymakers of structurally coherent territories have to respond reflexively to the new policies in order to retain that coherence. Where changes could no longer ensure coherence, a period of 'relative discontinuity' sets in, often triggering regulatory reconfiguration. State rescaling is therefore a change that re-defines national-level structural coherence; whether it disrupts or, perhaps counter-intuitively, reinforces this coherence has become a key research focal point on the periodisation of political–economic evolution. How relevant, then, are these insights for explaining changing regulatory dynamics in China?

Research has demonstrated how the gradual but ultimately expansionary exposure of the Chinese political economy to transnational capital has triggered the urbanisation of means of production (Ma 2005; Shen 2007; Li and Wu 2012). After the Deng Xiaoping government re-established cities as the primary sites of engagement with transnational capital, the platform was established for the emergence of a new inter-scalar rule regime that became increasingly dominated by growth-oriented urbanising initiatives (Tsing 2010; Wu 2015). And as introduced in Chapter 1, a series of 'nationally strategic new areas' have been designated in selected city-regions to smoothen and enhance global economic engagements through policy experimentation. As subnational policies and practices gain prominence over the highly-politicised, national-level 'permanent revolution' of the Mao era, state rescaling has unsurprisingly become an attractive concept to frame Chinese political-economic evolution. Scholars have reflected critically on the conceptualisation of state rescaling as a strategic response to

Fordist–Keynesianism, however, and a succinct overview of these critiques vis-à-vis the Chinese developmental trajectory demonstrates at least four reasons for developing a revised framework.

First, while state rescaling in China appears similarly as a response to crisis tendencies of preceding regimes, socioeconomic regulation after 1949 was not predicated on a Fordist mode of production and its corresponding territorial strategy, 'spatial Keynesianism'. At first glance, it is possible to argue that the predominance of national regulation in Fordist economies overlapped that of Maoist 'new China'. In tandem with the Fordist–Keynesian regulatory approach was a rolling series of economically nationalistic projects based on rural industrialisation. As introduced in Chapters 1 and 2, the vision was first actualised through the land redistribution to poor peasants between 1950 and 1952, and was further intensified through the reconfiguration of national state space into the People's Communes (based in the rural hinterland) and industrial work units (based in the cities). As neoliberalism set in motion the new international division of labour in the late 1970s, new projects of/for capital accumulation were launched in targeted cities in China (just as they were in western Europe). The urban bias of this developmental process across China took definitive form by the mid-1990s and, like the situation in western Europe and (to a lesser extent) North America, has been regularly reinforced since (Shen 2007; Wei 2013; Wu 2017).

Yet the empirical presupposition of spatial homogeneity in the Fordist heartlands before the launch of urban-oriented state rescaling appears too generic, as was the presupposition of socio-spatial egalitarianism in Mao-era regulatory logics prior to Deng Xiaoping's introduction of market-like rule in 1978. Spatial Keynesianism was constituted in practice by historically-specific and geographically-uneven conditions such as post-colonial populist politics and centralised state bureaucracy in Ireland (Breathnach 2010); actually-existing urban bias in a Canadian economy based significantly on natural resource extraction and staples exports (Ley and Hutton 1987; Hayter and Barnes 1990, 2001); or, as Kevin Cox (2009: 116) demonstrates, it was 'at best a very, very stunted creature' in the USA. The economic phenomena demarcating a definite change to post-Fordism is further complicated by research demonstrating the reconfiguration rather than the total disintegration of Keynesian spatial strategies (Davoudi 2009; Olesen and Richardson 2012; Tomlinson 2012). Herein lies an important question for framing Chinese state rescaling: was national-level egalitarianism under Mao more of an aspiration rather than an actual fact?

As Chapter 2 has shown, the Mao era was paradoxically characterised by institutionalised uneven development. Directly contrasting the equalisation objective of spatial Keynesianism, the Mao administration instituted economically nationalistic projects on the basis of the urban–rural dual structure, decentralised regulation in the rural People's Communes; and centrally-planned industrial production in urban areas. Socioeconomic disparities were consequently entrenched *in situ*. More importantly, the post-Mao policies to urbanise means of production – including labour power previously delimited to atomised rural communes – do not imply a new preference for urban-scale regulation; the Mao administration always privileged the urban, with 18% of Chinese citizens enjoying enhanced social benefits in Chinese cities (Whyte 1996; Zhang 1997; Bray 2005). As Laurence Ma (2005: 483) observes: 'Despite Mao's personal anti-urbanism during the Cultural Revolution (1966–1976), the urban scale has always been the preferred scale on the part of officials'. What happened in post-Mao China, then, was more accurately a qualitative shift in central–local regulatory relations: while municipal governments enjoyed greater decision-making autonomy in resource allocation and engagement with transnational circulatory capital, a one-track, historically sequential devolution of regulatory power from the national to the city-region did not take place.

Second, the retention of a unitary and hierarchical state structure in China suggests urban-oriented rescaling is not an historical inevitability. Specifically, it remains unclear what policy *alternatives* were available as policymakers in the Fordist heartlands decided to concentrate means of production in city-regions during and after the late 1970s. More empirical clarity is also needed on whether the major city-regions chose to implement the new urban locational policies, or whether there were specific socio-ideological objectives behind what appeared to an inevitable evolution towards city-regionalism. In the Chinese context, the central government's preference for territorially-targeted policy experimentation in the cities clearly demonstrates its inherent distrust of transnational circulatory capital. Furthermore, the emergent urban bias in industrialisation continues to be constituted by differentiated pricing controls on rural production, the enforced reconfiguration of rural land-use, and the unwillingness of city governments to offer social benefits to migrant rural workers (Kanbur and Zhang 1999; Bramall 2007; Webber 2012). For this reason, the conceptualisation of state rescaling in China needs to consider how the urban bias in domestic resource allocation and engagement with transnational circulatory capital is neither a stand-alone nor a permanent process. Rather, the orientation towards one scale of accumulation (the city-region) cannot be independent of rural (primitive) accumulation and social transformation. This further suggests it would be premature to frame the contextually-specific emergence of Chinese city-regions as *the* new state spaces of/for capital accumulation.

The third reason for reconfiguring the state rescaling framework is to highlight the role of politicisation. Research has demonstrated how the reconfiguration of state space is a contested process with no predetermined cause or outcome. As John Harrison (2012) shows, city-regionalism in England became a political strategy mobilised by actors in response to the-then Labour government's ambivalence towards devolution. Mustafa Bayırbağ's (2013: 1142) work on Turkey similarly reveals state rescaling as an effect of the interaction of 'rival hegemonic projects':

> State rescaling is not the product of a structural tension between central government policies and inevitable local responses. It is a conflict-ridden process where rival hegemonic projects clash as they strive to (re)define the meaning of 'nationhood' and 'national interest', which determine the (future) spatiality of exclusion from/ inclusion into national public policies. Then, a pro-decentralization agenda promoted by a (counter-)hegemonic project can be interpreted as a tactical move in the history of this struggle. Therefore, decentralization reforms cannot be readily associated with a particular type of a political project, or a particular historical period such as that of neoliberalism.

Bayırbağ's emphasis on 'structural tension' is particularly pertinent in the Chinese context, both during and after the Mao era. The Mao administration's quest to base 'China's tomorrow' on the Soviet Union regulatory blueprint was characterised by high profile intra-party conflicts, as exemplified by the sudden sacking of two senior revolutionary cadres, Gao Gang and Rao Shushi, in the mid-1950s; the public humiliation of reform-minded cadres like Liu Shaoqi, Deng Xiaoping, Xi Zhongxun and Bo Yibo during the Cultural Revolution (1966–1976); and the political intrigue surrounding the sudden demise of Mao's appointed successor, Lin Biao, in the early 1970s.

Yet, in a development unthinkable just a few years before, even Mao Zedong's regulatory approach mutated drastically from relatively localised experiments modelled after Soviet communes to more direct engagements in the 1970s with what the Chinese propagandistic machine termed 'American imperialism and its lackeys' (*mei diguozhuyi jiqizougou* 美帝国主义及其走狗). Particularly telling was the involvement of Chinese banks coordinating trade and raising capital in Hong Kong since the mid-1960s, ironically enabled by the circulation of Hong Kong dollars – a product and legacy of imperialism – throughout the Mao era (Schenk 2002, 2007). It appears the Mao administration had long adopted the 'yellow cat, black cat' pragmatism so popularly associated with Deng Xiaoping's market-oriented reforms in 1978 – any spatial strategy is a good strategy so long as it ensures regime stability.

Post-Mao attempts to institute a 'socialist market economy' were likewise unstable. In the late 1970s and early 1980s, Deng Xiaoping had to address opposition from conservative cadres like Hua Guofeng, Chen Yun and Yao Yilin, before negotiating divergences with reform-minded members of his administration like Hu Yaobang, Zhao Ziyang and Hu Qili. Perhaps the most prominent exemplars of marked intra-party differences in recent times were the public purges of Liu Zhijun, the former railway minister who ran the sprawling transport bureaucracy as a fiefdom, in 2011, and Bo Xilai, the Party Secretary of the Chongqing city-region who was associated with a series of law-bending 'socialistic' reforms, in 2012. Associated with each round of conflict were myriad socioeconomic institutions and political actors in different localities, each with their own agendas, interests and visions (ref. Wedeman 2003; Shih 2008; Ngo et al. 2017). In this respect, state rescaling in China exemplifies a recurrent, tension-filled and inter-scalar process that feeds back into what post-1978 Communist Party of China (CPC) regimes regularly terms the 'national strategy' of development (the pre-1978 term was the 'general line', a Soviet-inspired term that has not been officially repudiated).

Connected to this on-going attempt to actualise the 'national strategy' is the fourth reason why a geographical–historical contextualisation of state rescaling processes in post-Mao China is necessary. As introduced in Chapter 1, national-level political restructuring is entwined with place-specific policy experimenta-tion. This gradualist approach simultaneously encompasses urban and rural domains and differs from the marketising 'shock therapy' within the former Soviet 'socialist' bloc, as elaborated in Chapter 2. Rather, the lack of a 'unified national agenda' vis-à-vis increasingly differentiated subnational demands to engage with transnational circulatory capital has generated new impetuses for national-level regulatory adjustments. The challenge, then, is to develop an analytical framework that can explain and evaluate this dynamism.

Framing State Rescaling, Policy Experimentation and Institutional Path-dependency in China: A Dynamic Analytical Framework

As the foregoing section indicates, it would be problematic to describe or inter-pret the shifting institutional foundations of state regulation and socioeconomic development in China through a framework derived from western developmental experiences. This section will integrate the primary insight from western-based research – state rescaling as a strategic response to national-level regulatory failure – with two major aspects of Chinese political evolution: policy experimen-tation and institutional path-dependency. Expressed diagrammatically in Figure 3.1, this new analytical framework portrays national-level governance, which occupies a central position, as constituted by interactive tensions between four broad forces. These are namely (i) transnational circulatory capital searching for new locations of/for accumulation (ref. discussion of 'permanences' in the preceding section); (ii) subnational developmental agendas geared towards the political-economic goals of individual political actors; (iii) the quest by these actors to institute place-specific policy experimentation in the name of the 'national interest'; and (iv) the operational and ideological effects of inherited institutions.

Before exploring the dynamics and implications of these interactions, it would be helpful to introduce the following caveat: these four forces do not collectively comprise the sole origin and/or outcome of state rescaling in China. To be sure, there are different factors driving state rescaling and the framework presented is

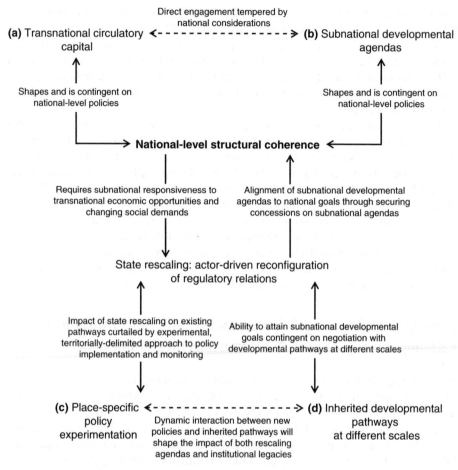

Figure 3.1   An integrated analytical framework on state rescaling in China.

necessarily partial. This said, these four forces were identified and integrated in the revised framework because they are major aspects of post-Mao politico-economic evolution. Unlike other East Asian 'developmental states', the CPC proactively engaged transnational capital to drive economic growth. Simultaneously, however, its retention of many Mao-era regulatory policies vis-à-vis the intensifying crisis tendencies within the global economy explains its tentative approach to private-driven marketisation (see, for instance, Horesh and Lim 2017). It is *within* this multi-dimensional context that this framework is developed; it is *of* this same context that this framework seeks a deeper and fuller understanding.

The Experimental Engagement with Transnational Circulatory Capital

Policy experimentation in growth-oriented projects is arguably the fundamental characteristic of national-level regulation in contemporary China. The content of experimental policies is not mandated in a top-down fashion; it is an outcome of negotiations between subnational actors (e.g. provincial secretary, municipal mayors) who are keen to involve transnational capital within their respective developmental agendas and central policymakers (both within the Politburo and the State Council) whose prerogative is to align differing agendas to political goals. In this regard, policy experimentation becomes a buffer zone within China's unitary and hierarchical political structure: some subnational actors compete with others for new regulatory powers to involve and integrate transnational capital *in exchange* for aligning endogenous plans with nationally-defined objectives (cf. Shirk 1993; Göbel 2011; Ahlers and Schubert 2015).

To retain control, the central government has been allowing potentially path-changing experimental policies to be launched on a 'move first, experiment first' (*xianxing, xianshi* 先行先试) basis. While former experimental policies have been extended nationwide if they prove successful, as correctly highlighted by Heilmann and Perry (2011) and Florini et al. (2012), they are increasingly retained *in situ* because of the need for geographically-differentiated engagements with transnational capital (Lim 2014). This competitive alignment to the previously-mentioned 'national strategy' is perhaps best summarised by the observations of Zhou Xiaochuan, the governor of the People's Bank of China (PBoC):

> The enthusiasm for reforms is very high at the local and grassroots levels, many provincial and urban governments and some organisations are earnestly requesting experimental grounds for reforms. Amongst the reasons is a common recognition that only through reforms would the consolidation of local economic development be possible, that it would be possible to push through various endogenous innovations and sustain local social stability. This point differs from some eastern European countries, in these countries a particular term known as 'reform fatigue syndrome' (*gaige pilao zheng* 改革疲劳症) has emerged, people are no longer motivated by or confident about reforms. In China, however, there is a lot of enthusiasm for reforms in all kinds of domains, proposals come incessantly from the grassroots level, in the hope that higher-level governments would allow them to launch experimental reforms. (Zhou 2012: n.p. Author's translation)

As Box 3.1 elaborates, subnational actors have immense incentives to lobby for and implement 'nationally strategic' policies. This corresponds with existing research that highlighted these actors as individuals seeking to advance their positions – formally and/or through clandestine means – within the existing administrative structure (Xu 2011; McGregor 2012; Pei 2016). This said, the

Box 3.1   Policy Experimentation and Central Power Preservation in post-Mao China

'Move First, Experiment First': The New Imperial Sword?

In imperial China, anyone bestowed the 'imperial sword' (*shangfang baojian* 尚方宝剑) would be a representative of imperial power and, concomitantly, possessed the authority to 'act first and answer afterwards' (*xianzhan houzou* 先斩后奏). In contemporary China, the institution to 'move first, experiment first' (*xianxing, xianshi* 先行先试) with so-called 'nationally strategic' policies has arguably become a new version of this 'imperial sword' – it is the object of much lobbying from provincial and municipal governments.

There are two reasons why this new institution is strongly coveted. First, the ability to launch innovative policies that move beyond from the standard parameters designated (ironically) by the central government allows immediate economic–geographical repositioning in relation to the global economy. Second, there is sufficient leeway for 'getting things wrong' (i.e. infringements that are inimical to national structural coherence). In other words, a government that can 'move first, experiment first' also has the 'power to be wrong' (*shicuoquan* 试错权). This power – which technically is an experimental *outcome* rather than conferred a priori as a legal right – has triggered widespread debates on whether it is antithetical to the rule of law. Unsurprisingly, the 'power to be wrong' is taken to be representative of a form of 'special power' (*tequan* 特权). Just like the power conferred on the 'imperial sword', the power to 'move first, experiment first' is taken as a directive to 'act first and answer afterwards'.

It is increasingly apparent that intention of the 'move first, experiment first' institution is to mediate the tensions associated with inherently unpredictable place-specific reforms and national-scale structural stability. In some ways this institution is integral to what Sebastian Heilmann and Elizabeth Perry (2011) term 'adaptive governance', namely the guerilla-like tendency for the CPC to break new regulatory ground spontaneously in different places before deciding whether to extend the new regulations nationwide (see discussion in Chapter 1). The gist of 'adaptive governance' is an underlying desire on the part of the CPC to transcend its existing governance structure without knowing exactly where this transcendence will lead. On the other hand, and this overlaps the key argument of this dissertation, the 'move first, experiment first' institution is integral to a process known as 'decentralisation as centralisation'. Through devolving the power to institute new policies to *selected* local governments, the central government enhances its leverage over uneven economic-geographical development. It first sets the targeted areas for policy reforms, and has to approve or veto the suggestions before they are implemented.

Through this leverage, the Chinese state apparatus could determine the extent to which its inherited institutions could be reformed without undermining the Four Cardinal Principles within the Chinese Constitution. And it is in this sense that the 'move first, experiment first' institution most resembles the 'imperial sword' – whatever happens on its travels, it remains in the first instance a function of the powers-that-be in Beijing.

primary research question in this book is not simply why these actors choose to promote their own agendas, but also why the central government is pursuing national-level change *through* some of these agendas.

Answers could be found through examining two interrelated dimensions of policy experimentation, namely (i) the *politics* (particularly the discursive justifications and counter-arguments) that led to the central government's eventual decision to launch (or negate) experimental policies in specific locations (cf. Li and Wu [2012]; Zeng [2015]); and (ii) the *interaction* between the experimental policies with national-level institutions inherited from the Mao era. As discussed in the preceding section, state rescaling occurs in response to regulatory failures. An emphasis on policy experimentation – now increasingly targeted at city-regions – as a *symptom* of potential regulatory failure offers new insights into aspects of the inherited institutions that subnational and central policymakers deem to have either failed or are almost obsolete. Through identifying these targeted aspects of change, state rescaling could foreground the constitutive – if not also constraining – effects of inherited institutions on national-level structural coherence (ref. Section 3.1).

For this reason, this chapter does not presuppose experimental changes to subnational pathways as antithetical to national-level institutional continuity. Some changes, such as the attempts by local cadres to institute 'nationally strategic' reforms in domains such as landownership and *hukou* in re-designated urban 'new areas', have been accepted insofar as they persuaded the Chinese central government of the potential to fortify national-level control (ref. Box 3.1). Other changes, such as the locally-driven initiatives in the late 1970s to institute household agricultural production and enable private trade, have been considered relevant because they triggered structural disruptions that consequently engendered new, if necessarily rudimentary, regulatory structures and accumulation regimes (more in Chapter 3). As Chapters 5 and 7 will show, the 'relevance' of any national-level institutional change, expressed through 'nationally strategic' policy experimentation in the Pearl River Delta and Chongqing, is to be assessed in relation to the 'coherence' of the prevailing regulatory structure, expressed through the ability of the Chinese central government to dictate and embed the flows and allocations of finance, production and labour power across the national scale. This emphasis on 'relevant changes' through policy experimentation does not presuppose the corresponding disruption of established regulatory paths – there is every possibility that change could be a precondition of continuity.

Cross-scalar Path-dependency and the Tensions of State Rescaling

Against the rolling series of territorially selective experimentation, an important challenge for research on Chinese state rescaling is to ascertain how these policies interact with other institutions not only across space but also *through time*. To follow Jessop et al. (2008: 392), 'sociospatial theory is most powerful when it (a) refers to historically specific geographies of social relations; and (b) explores contextual and historical variation in the structural coupling, strategic coordination, and forms of interconnection among the different dimensions of the latter'. Emphases are given to the notions of 'historically specific geographies' and 'contextual and historical variation' in this book's analytical framework because research increasingly present contemporary reforms in China as situated within inherited developmental pathways (ref. Chapter 1).

Path dependency has evolved into an increasingly unclear concept, however, and there is no systematic conceptual attempt in urban and regional studies to assess its relationship with socioeconomic reforms in China. To begin, it would be useful to define its parameters and connection with place-specific policy experimentation (ref. Figure 3.1). Arguably the most common definition of path dependence is the dependence of current and future actions/decisions on the outcomes of previous actions or decisions. As Scott Page (2006: 89) puts it, path dependence 'requires a build-up of behavioural routines, social connections, or cognitive structures around an institution'. Path formation is commonly construed as an accidental outcome; a chance event. Central to this process is the eventual formation of institutional 'lock in', whereby a practice or policy becomes effective or feasible because a large number of people have adopted or become used to this practice or policy. Any drastic alterations to the path, even in the face of inherently superior alternatives, would thus encounter resistance from groups of 'locked in' actors whose interests would be compromised by the proposed changes.

How policy experimentation interacts with inherited developmental pathways in and through the re-scaled 'nationally strategic new areas' offers a unique prism to ascertain the impact of regulatory reconfiguration on national-level coherence. At one level, path dependency is an attractive concept to explain how attempts to 'move first, experiment first', driven through the rescaling agendas of domestic political actors based within a range of subnational scales, interact with established institutions at different levels. In exchange for retaining some Mao-era institutions, these actors gained more 'freedom' to accumulate capital, consequently deepening their dependence on these institutions despite their limitations.

At another level, however, the conceptual application of 'path dependency' should be mindful of some of its biggest problems. One major problem is the lack of emphasis on the 'build up', in Page's (2006) parlance, to the formation of path-setting institutions. Developing this point, Guy Peters et al. (2005) argue that there exists a tendency in research on institutional path-dependence to accord history a logical trajectory, or 'retrospective rationality', such that available alternatives and political conflicts that occurred alongside more 'visible' historical processes are neglected. It is important, argue Peters et al. (2005: 1282), to be cognizant 'that prediction of persistence does not help at all in understanding institutional change'.

The notion of a 'logical trajectory' is further complicated by *geographical variegations* in the developmental pathways within nation-states. As Ron Martin and Peter Sunley (Martin and Sunley 2006; see also Martin 2010) have shown, the developmental paths of subnational regions are neither unique nor delimited to those regions. These studies correspond with Peck's (2002: 340) observation that 'the present scalar location of a given regulatory process is neither natural nor inevitable, but instead reflects an outcome of past political conflicts and compromises'. In this regard, the existence of a scale of socioeconomic regulation such as 'the city', 'the province' or 'the national' cannot be assessed narrowly from a singular spatiotemporal vantage point; a robust historicisation of *cross-scalar relations* – specifically the impact of policies instituted at one scale on other scales – is necessary.

This cross-scalar focus foregrounds two interrelated blind spots in the historical institutionalist literature on path-dependency: the *politics* that produced a specific 'path' is often unclear, as is the connection between institutional reforms at the subnational or supranational scales and national-level structural coherence. The possibility of institutional continuity-in-change through state rescaling suggests rescaling is not a linear historical process that is exclusively derived from market-oriented reforms instituted in and after 1978 (ref. preceding section). Rather, the post-Mao party-state has been working at various levels – albeit on a tentative basis – to drive development through reconfiguring Mao-era policies. Generated by and expressed through policy experimentation in targeted territories (e.g. marketising land use through 'land tickets' or *dipiao* (地票) in Chongqing and financial innovation in the Pearl River Delta city-region), the reconfiguration process illustrates the relevance of inherited policies at one scale (national-level path dependency) but also develops fresh regulatory capacities at another (local level path generation).

A New Research Agenda

Building on the foregoing emphasis on institutional continuity-and-change, this book implements a new research agenda that aims to make two advances. First, it moved beyond the relatively static periodisation that underpins state rescaling research by focusing on *tensions* between continuity (retention of inherited policies) and change (policy experimentation). In so doing, the book does not establish the existence of historical periods distinguished by specific regulatory scales (e.g. a Mao-era defined by national-level regulation, as discussed in Chapter 2); it seeks, rather, to establish the *historical significance* of contemporary change as pursued through state rescaling (e.g. *hukou* reforms in Chongqing). Second, it engages explicitly with the connections between scalar configuration and national-level coherence. State rescaling may have led to the privileging of specific subnational scales (particularly the city-region), but this is arguably a means to attain a broader objective – the sustained stability of the national structure. The agenda is underpinned by four interrelated questions:

(a)   If actors positioned at multiple scales proactively clamour for place-specific policy experimentation after 1978, does it mean only centrally-driven regulation was previously predominant under the Mao Zedong regime (1949–1976)?

(b)   Could pre-1978 China be periodised as a structurally coherent, nationally-oriented and centrally-driven developmental approach such as 'spatial Keynesianism'?

(c)   What regulatory crises triggered successive waves of state rescaling in China? What aspects of regulatory failure have central policymakers been responding to as they accept proposals to 'scale up' or 'scale down' socioeconomic regulation?

(d)   What does the success and/or failure of policy experimentation in the targeted territories reveal about national-level institutional continuity and change?

These questions co-constituted the research design of this book. The research process began with the observation and documentation of experimental policies introduced in selected city-regions (questions (a) and (b); ref. Chapter 1). Here, specific questions were raised regarding the content and applicability of the new policies. As mentioned earlier in this section, this was set within an actor-focused analysis that evaluated the specific agendas and discourses of the primary proponents of these reforms. To facilitate a clearer comparison, research was also conducted on the preceding policies and their accompanying discourses, which leads to the subsequent part of the research. This segment aimed to re-evaluate how inherited institutions enabled, guided, channelled and constrained policy experimentation across post-Mao China (questions (c) and (d)). Each targeted territory of/for policy experimentation contained its own set of spatio-temporal relations with actors and institutions, and it was through identifying these relations that made it possible to ascertain whether reforms are truly transformative.

Through this multi-layered, mutually-reinforcing attempt to evaluate the past from the lens of the present, and to conceptualise the present through ascertaining the impacts of policies inherited from past regimes, the framework problematises simple 'transition' models that portray a unidirectional, epochal change in the post-1978 Chinese political economy, a change characterised by decentralised governance and intensified economic-geographical inequality (as discussed in Chapter 2). It emphasises, instead, a more deeply sedimented pattern of development that was (and remains) marked simultaneously by significant (and enduring) forms of uneven socioeconomic development and experimental (and capricious) attempts to transcend them.

Conclusion

State spatial reconfiguration is a central feature of socioeconomic reforms in post-Mao China. This process is unfolding against a dynamic context of economic globalisation, the corresponding rise of neoliberalism after the 1970s, and the CPC's insistence on retaining a strong state role in socioeconomic regulation. While a growing scholarship has defined this restructuring as a 'scaling down' from national-level central planning to urban entrepreneurialism, what remains unclear is its connection to research showing the CPC to be more robust, resilient and flexible (Nathan 2003; Zeng 2016). By extension, how state rescaling reproduces – if not also reinforces – the national scale as a regulatory platform has not been explicitly addressed. This chapter is an attempt to foreground and address this conceptual gap.

Working from a critical review of conceptual and empirical work on state rescaling, the chapter highlighted four main reasons why a recalibration of the original, western-focused framework is necessary. First, the geographical–historical conditions that generated state rescaling across China differed fundamentally from those associated with the crisis of Fordist–Keynesianism. Second, socioeconomic reforms do not inevitably privilege the city-region as the primary regulatory scale. This ties in with the third reason: the politicisation of state rescaling involving specific interest groups has been relatively overlooked. Last, but not least, the Chinese party-state continues to prize major institutional foundations established during the Mao era, which explains its preference for territorially-contained policy experimentation rather than 'big bang', national-level reforms. These reasons collectively underpinned the development of a new framework and research agenda that illustrated the co-constitutive relationship between successive rounds of policy experimentation and institutional renewal since the CPC took political power in 1949.

At one level, the framework offers a platform to examine the connections between institutionally-distinct locations that are jointly pursuing 'nationally strategic' reforms (e.g. the Guangdong Free Trade Zone vis-à-vis the Shanghai Free Trade Zone). As Brenner (2009: 42–43) puts it, the 'institutional and spatial coherence' of a scale 'can be grasped only with reference to their distinctive roles and positions within interscalar hierarchies'. Ascertaining the relations of place would thereby entail tracing the constitutive extra-local processes of an apparently localised phenomenon (e.g. how the launch of cross-border flows of the Chinese currency – the renminbi [RMB] – between Shenzhen and the Special Administrative Regions [SAR] of Hong Kong and Macau as a three-way process involving the central, provincial, and the SAR governments).

At another level, the framework facilitates a robust historical engagement that went beyond periodising the entire Chinese political economy. Through tracing the historical significance of experimental policies in individual locations, the framework avoids treating political economies as internally homogeneous over time. Rather, the path-dependency of specific rescaling processes was called in question. Cognisant of the contributions and constraints of the path-dependency paradigm, the framework situated state rescaling as simultaneously a *reaction* to and an attempt to transform inherited institutions. In so doing, it circumvented the pre- and post-Mao temporal dichotomy and left open-ended the interpretations of policy experimentation in selected city-regions. Specifically, contemporary rescaling processes became empirical platforms for historical re-evaluation. To re-borrow Jessop's (2001) terms, if the Mao-era regulatory 'structure' remains characterised by 'coherence' (and this 'structure' includes the spatial configuration of socioeconomic activities), the post-1978 changes triggered by Deng and his successors would be relevant insofar as they ensured and extended the stability of this 'structure'. The implication of this 'relevance' is clear: 'liberalisation' reforms, expressed through increasingly differentiated experimentation in targeted territories, have thus far been a function of a 'transition' towards an as-yet-determined end-state. A major challenge for urban and regional research, then, is to ascertain the extent to which these reforms, each increasingly taking on distinct city-regional forms, are truly disruptive of the existing structure, or whether change has become a means to reinforce national objectives already concretised in the heady days of the 'socialist high tide'.

Part III
State Rescaling in the Pearl River Delta and Chongqing

Chapter Four
Becoming 'More Special than Special' I
The Pressures and Opportunities for Change in Guangdong

Introduction

The Greater Pearl River Delta (hereafter GPRD) region in the south-eastern corner of China has functioned as the national economic 'motor' since experimentation with global economic (re)integration began in 1978. By official definition, the GPRD comprises nine cities and two Special Administrative Regions (SARs, namely Hong Kong and Macau). Encompassing less than 1% of China's total land area and registering less than 4% of the total national population, the nine cities in the GPRD account for almost 10% of the country's GDP[1] over the last decade; during the same period, the region embedded an estimated 20% of its foreign direct investment (FDI) and generated approximately 25% of national trade. The location of three of China's first four Special Economic Zones (SEZs), namely Shantou, Shenzhen, and Zhuhai, the region is part of a broader Guangdong province that was identified as the destination for the largest number of domestic migrant workers in the 2010 Chinese Population Census. Viewed as an integrated whole, this extended, cross-border metropolitan region is now deeply articulated in the global system of capitalism. With the two SARs functioning as 'free' hubs of/ for capital flows and the other nine cities converting these flows into industrial production (through which invested capital would then be valorised when the finished products are re-exported), the GPRD is arguably the region in China where the fixities and flows of capital have been most intense over the past three decades.

On Shifting Foundations: State Rescaling, Policy Experimentation and Economic Restructuring in Post-1949 China, First Edition. Kean Fan Lim.

Going by the emergent economic–geographical changes in the GPRD, there is no letting up in the quest to capture new investments. Between 2009 and 2012, the Chinese central government approved the launch of three inter-related 'new areas', each occupying a corner of the Pearl River Delta (PRD). As Figure 4.1 shows, Hengqin New Area is located on the western corner of the PRD, just adjacent to Macau SAR; Qianhai New Area is to the northwest of Hong Kong, just under 30 minutes of land connection following the completion of a new railway; and Nansha New Area, the newest of the three, is located to the northern tip of the PRD, directly contiguous with Guangzhou, the capital of Guangdong province.

Exemplifying a new wave of state rescaling, defined in Chapter 1 as a reconfiguration of regulatory relations between governments across state space, the official objective of these three zones is to deepen economic integration with the Hong Kong and Macau SARs through a form of 'staggered development' (*cuowei fazhan* 错位发展). Together with infrastructural investments to enhance the time–space compression of the PRD, the development of the three 'new areas' constitute a seven-project list in the Chinese government's 12th Five-Year Plan to effect

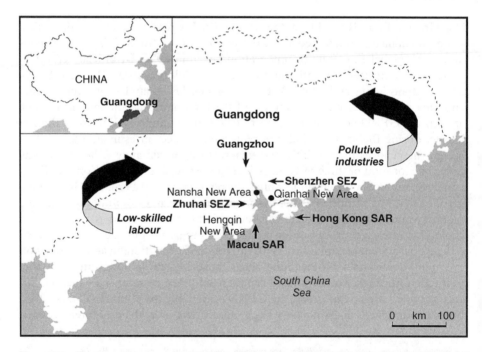

Figure 4.1   New economic–geographical transformations in the greater Pearl River Delta: Hengqin, Qianhai and Nansha new areas and the relocation (in block arrows) of two unwanted 'highs': the high number of low-skilled labour and pollutive industries in the GPRD. Source: Author, with cartographic assistance by Elaine Watts.

a broader spatial project known as the 'Deepening of Guangdong-Hong Kong-Macau Co-operation'.[2] As the-then Party Secretary of Guangdong province, Wang Yang, puts it, this economic–geographical reconfiguration would actively alter the spatial relations of the entire GPRD:

> Amongst these there is the Hong Kong-Zhuhai-Macau Bridge, the Hong Kong-Shenzhen-Guangzhou direct train link, these [treasures] involve seven major projects within one province, all these are of our province, this is what other provinces absolutely do not have, many provinces cannot even make one mention of this term [i.e. 'treasure']…The outline of the 12th 5-Year Plan provides a rare opportunity for regional cooperation, Guangzhou's Nansha is included in the outline, the development of Qianhai in Shenzhen is included in the outline, Zhuhai's Hengqin is in the outline too, this is like three chess pieces put together, these three chess pieces are three important nodal points in Guangdong's industrial transformation and upgrading. (*Zhuhai Daily*, 8 March 2011)

This two-chapter segment shows how the designation of these 'nationally strategic new areas' in Guangdong simultaneously facilitated place-specific policy experimentation and expressed the constraints of institutional path-dependency (ref. framework presented in Chapter 2). As Chapters 1 and 2 have demonstrated, the overarching spatial logic of socioeconomic regulation was to produce and retain an integrated national political economy. This goal became increasingly challenged over the last two decades as global economic integration deepened. The biggest difficulty of the reforms in Guangdong was thus to ensure sustained economic growth without unsettling the objectives of macroeconomic regulation. For this reason, the attempt to reconfigure Guangdong's economic geographies – and by extension regulatory relations between the central, provincial and city governments – was a distinct political strategy that in turn reflects the dynamism and uncertainty of state rescaling.

As the empirical research presented in this chapter reveals, there was no immediate tendency for firms, especially the medium and small enterprises that comprise the backbone of Guangdong's intricate production networks, to relocate en masse from the Pearl River Delta to less developed regions in the province.[3] Even when the global financial crisis struck in 2008, firms were not the primary agents clamouring for change. Yet, the global financial crisis provided an opportune backdrop for the Guangdong government to precipitate a 'double relocation' (*shuang zhuanyi* 双转移) industrial policy. The first relocation involves shifting labour categorised as 'low-skilled' and firms categorised as 'high in pollution, high in energy use and low in efficiency' (*lianggao yidi* 两高一低) from the core Pearl River Delta region to the underdeveloped regions of the province.[4] Effectively a state-driven form of value chain upgrading, this policy in turn set the platform for a simultaneous inflow of advanced services and higher-order manufacturing. To jumpstart these inflows, 'nationally strategic' reforms were introduced in Hengqin, Qianhai and Nansha under the 'move first' experiment first' institution (ref. Chapter 1; specific impacts of policy experimentation to be analysed in Chapter 5).

Against this backdrop, this chapter makes the argument that the emergence of Hengqin and Qianhai New Areas (and, by extension, the GFTZ) must be viewed within the broader framework of state rescaling, policy experimentation and path-dependency. It develops the point raised in Chapter 3 that state rescaling in post-Mao China comprise a form of 'adaptive governance', through which actors positioned at different governmental levels (re)negotiate regulatory relations. Just like Deng's SEZs were not part of a 'unified national agenda', these new 'nationally strategic' sites were not part of a unified 'national strategy' (ref. Chapter 1). Neither could this emergence be reduced to cost-based calculations internal to firms (cf. Yang 2012). In fact, as Wang acknowledged in a May 2009 discussion, firms were quite resistant to the 'double relocation' policy:

> Some people felt life was quite good, so long as there was money to be earned, why force issues? Some felt industrial upgrading is a long-term process, was it necessary to trigger this through industrial relocation? And a third type was of the opinion that industrial relocation would have a direct impact on the local economy, especially at the town and village levels, hence their strong opposition. (*Nanfang Zhoumo*, 13 August 2009; author's translation)

It is thus apparent that the 'double relocation' policy was integral to Wang's agenda to secure political results (*zhengji gongcheng* 政绩工程). Problems emerged when the relocation policy was perceived to be undermining national structural coherence. This consequently foregrounded two contradictory aspects of the relationship between the Guangdong provincial government and central policy-makers in Beijing. At one level, the 'double relocation' policy reflects a specific post-Mao 'political logic' of reforms known as 'reciprocal accountability' (Shirk 1993). This logic invites developmental initiatives of national significance from local governments, in the anticipation that these governments would reciprocate by aligning their initiatives to national objectives (ref. review in Chapters 1 and 3).

At another level, however, the policy paradoxically (i) goes against the 'national chessboard' pathway and, simultaneously, (ii) undermines the newer industrialisation pathway along the eastern seaboard instituted during the 7th Five-Year Plan (1986–1990). In this regard, Guangdong policy-makers' attempt to territorially reconfigure the provincial industrial composition – and, by extension, regulatory relations between the central, provincial and city governments – more accurately reflects tensions with national-level regulatory strategies launched by earlier regimes. Something needed to be done, then, to ease the negative impacts on GDP and employment, and this took the form of successive 'nationally strategic' reforms in Hengqin, Qianhai and Nansha New Areas.

This chapter will comprise four parts. The debates on and politics surrounding the 'double relocation' industrial policy will be presented in the following section. The section demonstrates how these central-provincial politics is embedded within the broader geographical-historical context of the GPRD. As Brenner (2004: 81) notes, 'state scalar configurations must be conceptualised in a manner that is explicitly attuned to the historicity, and thus the malleability, of each scale of state institutional organisation, regulatory activity, and political struggle.' The third section then evaluates the tensions generated by the 'double relocation' policy. It showcases how simultaneous industrial relocation and value chain upgrading was viewed as an (im)possible task. It was for this reason that economic restructuring in Guangdong became a *national political issue*. As the fourth section will elaborate, this issue consequently generated the driving force to 'scale up' the territories of Hengqin, Qianhai and Nansha into 'nationally strategic new areas'. The concluding section emphasises how state rescaling in Guangdong was contingent on the economic–geographical reconfigurations in ways that were never ascertained a priori by the Chinese central government. Conceptually, this phenomenon brings into question whether state rescaling has become a *function* of national-level structural coherence (as discussed in Chapter 3; ref. Figure 3.1), or, perhaps counter-intuitively, whether it is a new tool for local governments to perpetuate the economistic approach to development known as 'GDP-ism'.

The Economic–Geographical Backdrop: 'Double Relocation' in Post-crisis Guangdong

The development and emergence of Hengqin and Qianhai New Areas are inextricably entwined with the evolution and impacts of spatial projects implemented at the provincial and national level. For the large part of the past three decades, Guangdong had maintained its leading position in the Chinese economic–geographical hierarchy because of preferential policies launched during the Deng era. This position in turn expresses a systemic coastal bias that saw provincial and city governments along the more industrialised eastern seaboard enjoy preferential treatment from central government agencies vis-à-vis their counterparts in the less-developed western and north-eastern interior. As presented in Chapter 1, Deng Xiaoping's macro-scale approach to spatial reconfiguration was based on the 'ladder-step theory'. This prescriptive 'theory' – more accurately defined as a policy blueprint – delineated Chinese state spatiality into three economic belts: the eastern (coastal), central, and western. Coastal cities and provinces were given the priority to ascend the development 'ladder', on the proviso that capital accumulated from the 'first mover' belt would diffuse downwards to other rungs of the ladder (see Fan 1995; Wang and Hu 1999). Yet it was never clear when this diffusion would occur; in China today, there are increasing worries that a spatial equalisation of living standards would never happen (Mao 2013).

Because of Deng's macro-level spatial project, Guangdong ranked as China's top province by GDP annually since 1989. Statistics for 2007 indicate that 96% of import/export trade in Guangdong was concentrated in the PRD, which suggests economic data for Guangdong province would be an effective measure of the economic performance of the PRD (Guangdong Bureau of Statistics 2008). Placed in relation to national-level import/export figures, Guangdong on average accounts for almost 30% of the national total since China's accession to the WTO in 2001 (National Bureau of Statistics 2013). Along with the growing inflows of capital, Guangdong province – and the PRD in particular – has been the top destination for China's expanding 'floating population' of migrant workers (*liudong renkou* 流动人口) in the last decade. According to the Chinese national population census of 2010, a third of the country's 'floating population' of around 200 million (this figure has since increased to 236 million in 2012) was concentrated in Guangdong. From these statistics on trade and employment, it can be inferred that, across Chinese state space, the PRD is by far the most deeply 'articulated' of the Chinese city-regions into the global system of capitalism.

The economic–geographical reconfiguration of Guangdong and its multidimensional relationship with foreign capital is a snapshot of the CPC's post-1978 politico-economic reforms (ref. Chapters 1 and 2). Within Guangdong, the dominant role played by foreign capital allowed GDP to grow without the concomitant emergence of a large private capitalist class that is capable of undercutting CPC interests (many of the manufacturing subcontractors are in fact medium and small enterprises). With Hong Kong and Macau SARs functioning as geographically-contiguous offshore conduits for the valorisation and reinvestment of financial capital, this strategic, state-driven engagement with foreign capital produced what Lin (1997) calls 'red capitalism in south China'.

As with all processes, the deep articulation of Guangdong-based enterprises and labour power in global production networks generated its own developmental pathways. And at the onset of 2008 global financial crisis, this pathway became an economic–geographical liability. For the whole of Guangdong province, the trade dependency (as % of GDP) was 130% in 2007. By comparison, the national average was (an already-high) 66.2% in 2007. Towards the end of 2008, it was clear that the global financial crisis precipitated a drop in effective global demand for manufactured goods from Guangdong-based industries. The province's total volume of imports and exports decreased sharply for eight consecutive months from November 2008, with the largest monthly decline rate reaching 31.1%. Foreign capital investments dropped by over 50%.

According to customs statistics, the total import/export value in Guangdong recorded 257.87 billion *yuan* (~US$40.1 billion) in the first half of 2009, a 20.7% decrease year-on-year (*People's Daily*, 3 August 2009). Through an investigation spanning three years, the *Nanfang Dushibao* (2 April 2012a) reported that two waves of factory closures followed after the crisis struck. Between 2008 and 2009, it was estimated that half of the 58,500 Hong Kong-owned export-processing subsidiaries would not survive, while a new wave of closures affecting more sectors ensued in 2011. Quite clearly, a strong exposure to international trade had become a double-edged sword for the Guangdong economy.

In a candid evaluation published in 'The Outline of the Reform and Development Plan for the Pearl River Delta (2008–2020)',[5] the Guangdong government delineated how earlier approaches to capital accumulation in the PRD have generated their own internal contradictions:

- The overall industrial body is of the lower-order; value-added of product is not high, the trade structure is not reasonable; innovative capacities are insufficient; overall competitiveness is not strong.
- The degree of land development is excessive; the ability to conserve energy and resources is relatively weak; problems with environmental pollution are relatively pronounced; resource limitations have become apparent; hence the traditional model of development cannot be sustained.
- Urban–rural and regional development remains uneven; the allocation of productive forces is still not reasonable; the efficiency of spatial usage is not high.
- Social projects are relatively lagging behind; the development of human resources, public service standards and 'soft' cultural capabilities are to be improved.
- The administrative management systems, social management system and other areas continue to face heavy reform tasks; tackling difficulties of reform have become greater.

Wang Yang responded to these contradictions with a strong dose of 'shock therapy' almost immediately after moving to Guangdong in December 2007. This response first took shape in March 2008 through four instructions to the local government of Dongguan, Guangdong's major manufacturing hub located between Shenzhen and Guangzhou (the provincial capital). The instructions were, namely, to (i) push through the readjustment of the industrial structure and the 'transformation and upgrading' (zhuanxing shengji 转型升级)[6] of commodities produced; (ii) reduce the urban population and improve labour force quality; (iii) engage in comprehensive planning and gain momentum in the restructuring process; and (iv) remain resolute in response to challenges, with a view of dismantling rigid ways of thinking, rigid developmental approaches, and the solidified constellation of interest groups (Guangzhou Daily, 27 March 2008a).

It is important to note that the global financial crisis was still not full-blown when Wang delivered his instructions. This strongly suggests industrial reconfiguration in Guangdong was part of a developmental agenda independent of the crisis. The primary goal of these instructions was to enable Guangdong to break out of the established path of export-oriented production. Wang summed up his approach through a strong warning to the Dongguan cadres: 'If Dongguan does not reconfigure its industrial structure today, it will be reconfigured by the industrial structure tomorrow' (*Guangzhou Daily*, 27 March 2008a). Interestingly, this description expresses the *necessity* of subnational socioeconomic reconfiguration vis-à-vis transnational circulatory capital (ref. Figure 3.1, Chapter 3). Through asserting Dongguan would be 'reconfigured by the industrial structure tomorrow', Wang demonstrated an awareness that firms would eventually relocate to places that could offer higher rates of profits. And as a major global manufacturing site for TNCs, the Pearl River Delta was especially vulnerable to these pressures.

To enable firms to generate higher profit rates, Wang believes that places possess the agency to pre-empt the effects of spatial divisions of labour by reconfiguring the industrial structure. The embodiment of this agency is the ability of the local governments to influence, in a proactive and strategic manner, the allocation of capital. Attaining this ability would require policymakers to quickly identify falling rates of profit and/or growing economic inefficiencies in firms within their respective jurisdictions. To this end, Wang implored Dongguan's cadres, the 'government must fully develop its impact by devising and implementing policies for industrial reconfiguration, transformation and upgrading' (*Guangzhou Daily*, 27 March 2008a).

New changes began to unfold rapidly thereafter. In May 2008, less than two months after Wang's visit to Dongguan, the Guangdong government released a province-wide economic restructuring plan known as 'Decisions on pushing forth industrial and labour power relocation' (hereafter 'Decisions'). Central to the 'Decisions' is the economic–geographical strategy of 'double relocation' (*shuang zhuanyi* 双转移). As mentioned earlier in this chapter, these 'double relocations' refer to relocating targeted enterprises and the labour power they employ from the PRD to 'relocation industrial parks' (*chanye zhuanyi gongyeyuan* 产业转移工业园) in the less developed regions of the province, namely the eastern and western ends and the northern highlands (ref. Figure 4.1). These targeted firms and labour were those generating 'two highs and one low', i.e. high in pollution, high in energy use and low in efficiency.

In turn, a parallel strategy to bring in advanced manufacturing and/or higher-order services would be introduced to enhance the industrial composition of the Pearl River Delta. Wang offered a detailed justification of his government's restructuring rationale through the CPC's mouthpiece, *People's Daily* (17 October 2008b; author's translation):

To create an innovative Guangdong model, it is necessary to grasp key domains for autonomous innovation and pivotal sectors and work to overcome the technological barriers that constrain socioeconomic development in Guangdong. While there is a need to enhance the innovative tendencies within traditional industries experiencing transformation and upgrading, there is also a need to accelerate the industrialization of innovative technologies; while there is a need to push though new up-and-coming industries, there is also a need to implement technological projects amongst privately-owned sectors, so as to ensure the fruits of innovation are more widely-shared.

Labelling the industrial upgrading project 'emptying the cage to change the birds' (*tenglong huanniao* 腾笼换鸟), Wang emphasised that his aim was to preclude economic 'hollowing out' by encouraging the growth of more high-tech firms:

> For the Pearl River Delta, the point is to develop higher-order technologies endogenously, within the productive mechanisms of traditional industries, in turn realizing the true meaning of 'emptying the cage and changing the birds' (*tenglong huanniao*), or it could be called 'expanding the cage and strengthening the birds' (*kuolong zhuangniao*). On this basis, the situation of economic 'hollowing out' in the process of industrial transformation and upgrading could also be avoided. (*People's Daily*, 17 October 2008b; author's translation)

Wang's essay was clearly an attempt to justify the suddenness and extent of his developmental agenda after its implementation sparked widespread discussion and debate over its political and economic implications. In a Chinese political circle where key actors predominantly deploy codified language 'internal to the institution' (*tizhinei* 体制内) to convey their feelings, latent meanings are often expressed through symbolic metaphors. Viewed in historical perspective, Wang's metaphorical choice of 'cage' and 'birds' appears deliberate. The original user of these metaphorical terms was Chen Yun, the previously mentioned economic advisor to Mao Zedong and, during the post-1978 reform era, an opponent of Deng Xiaoping's (relatively) liberal approach to economic governance. Chen argued that the state could not hold 'birds' (i.e. capitalistic actors) tightly in its hands because they would suffocate; yet if the grip on the 'birds' loosened, they would inevitably fly away. The middle ground is to construct a sturdy 'bird cage' (i.e. an economy strongly regulated by the everlasting arms of the CPC) that allows the 'birds' to fly and breathe, but only within the parameters of the 'cage'. Widely-known in China as 'birdcage economics' (*niaolong jingji* 鸟笼经济), Chen's logic arguably underscores the logic internal to the CPC's overall approach on socioeconomic regulation today: giving more 'decisive' roles to 'the market' proceeds *in tandem* with reinforcing state economic involvement (cf. Lin 2011).

Wang Yang's extension of Chen Yun's metaphor to popularise his province-level restructuring project indicates institutional path-dependency in at least two instances. First, in an affirmation of the Four Cardinal Principles (ref. Chapter 1), it re-emphasised the necessity of the CPC to determine the geographical parameters for economic 'birds' seeking to manoeuvre within Chinese state space. The economic relations of the Guangdong regulatory regime across and beyond China were thus literally reconfigured through industrial policies that reconfigured spaces of production. That Wang was advocating proactive state intervention goes against conventional interpretations of his role as a (neo)liberal reformer. Indeed, as shall be shown shortly, Wang's restructuring project could take place only because he had the support of the policymakers at the highest echelon – specifically, because he had the support of then Chinese President Hu Jintao.[7] Second, in seeking to retain the economic competitiveness of the PRD through significant economic–geographical reconfiguration, Wang was making clear his intention to stay on the top rung of Deng's 'ladder step' model. As mentioned earlier, this model presumes a 'levelling out' in living standards would occur automatically. Prior to the 2008 'Decisions', Deng's presumption had become the status quo of economic–geographical evolution in China; no concrete attempt had been launched to reconfigure the national economic–geographical structure (the evolution of Deng's strategy was discussed in Chapter 1). From Wang's subsequent reflections, it can be seen that his fundamental intent in Guangdong was to ensure institutional continuity *through* change:

> Practice has proven that with the conformity to market rules; with the courage to break through the path dependency that encumbers development; with the elimination of unreasonable interest groups; with the timely increase in developmental quality; and with putting the ability to sustain development in first place, it is definitely possible to create a developmental model that could deliver high value-added products, good quality development and personal income growth that is commensurate with the growth of the economy. (*People's Daily*, 24 August 2012)

Read in relation to his imploration to Dongguan cadres in March 2008, Wang's use of 'market rules' (*shichang guilü* 市场规律) to justify breaking out of the labour-intensive, export-oriented 'path dependency' (*lujing yilai* 路径依赖, the exact equivalent of the term Wang employed) only serves to reinforce the importance of state intervention in the regulation of the Chinese political economy. As he made clear during the Dongguan visit, the reconfiguration of spatial divisions of labour could – and should – involve proactive action on the part of local governments (ref. *Guangzhou Daily*, 27 March 2008a). Furthermore, a clause in 'Decisions' specified that the restructuring process required a fiscal injection of 50 billion yuan (~US$7.81 billion) over five years. Even with this financial support, there was no guarantee that PRD-based firms – should they not have shut down at all, as have happened to the purported 50,000 firms at the time – would be willing to relocate to less developed regions within Guangdong province. This essentially means the Guangdong restructuring plan – discursively justified as following 'market rules' – was (and remains) an experimental project that entailed the state to underwrite the financial risks. Viewed in relation to the discussion on state-directed uneven development in Chapter 2, Wang's approach underscores how the reconfiguration of regulatory relations does not occur because of 'market rules'. As the next section will elaborate, shifting regulatory relations across scales constitute a function and an outcome of central-local politics.

On the (In)commensurability of Spatial Restructuring and Economic Growth

Two contradictions were intrinsic in Wang's ambitious approach to shape up and ship out industrial actors within the Pearl River Delta at the same time. At one level, it is clear that the pursuit of new competitiveness in Guangdong, which perpetuates the coastal bias in development, contradicts Deng's 1988 vision of attaining similar developmental standards across the country. As discussed in Chapter 1, Deng pledged to even out income disparities over time. In approving new initiatives for to enhance economic–geographical advantage in the Pearl River Delta over other parts of China, however, the Chinese central government was effectively entrenching the pre-existing uneven development. In a sense, it could be argued that the central government was 'locked in' to its earlier choice to privilege Guangdong in its industrialisation policies. As the host of industries that employ the majority of the domestic 'floating population' (ref. Chapter 2), it would not be easy to allow deindustrialisation to occur in Guangdong without first establishing alternative employment opportunities elsewhere. While attempts were launched at the same time in interior China, spearheaded by the rescaling of Liangjiang New Area in Chongqing, they were arguably longer-term solutions (ref. Chapters 6 and 7).

The conundrum for national policymakers was how to bring in 'upgraded' industries while simultaneously relocating undesirable industries from the PRD. It must be noted that this injunction in 'Decisions' was in itself an experiment: 'emptying the cage' was not only an intra-provincial issue, it would generate direct impacts on national employment given Guangdong's position as the top destination for migrant workers. The primary goal of this injunction, as Wang made clear through the *People's Daily* in October 2008, was to preclude industrial 'hollowing out' (see preceding section). While theoretically sound, it was unclear how the injunction could be put in practice: firms began to shut down and replacements were hard to find in a period of immense economic uncertainty. Even if high-end industries were willing to move into the PRD, emerging empirical evidence at the time indicated it would occur only after a time lag. It was for this reason, as shall be elaborated shortly, that the central government became concerned with the reconfiguration process. How, in the midst of economic slowdown, could the local governments across the province 'strongly develop advanced manufacturing industries, high-order advanced technological industries, modern services and equivalent high value-added industries' ('Decisions', n.p.; author's translation)?

Within two months of the implementation of 'Decisions', there was widespread speculation that the local government in the manufacturing hub of Dongguan had begun 'driving away factories, driving away people' (*ganchang ganren* 赶厂赶人). Local cadres purportedly went about 'encouraging' enterprises to relocate. The industrial 'hollowing out' was arguably intensified by two developments, namely (i) the claim of one Dongguan cadre that discussions were underway to halve the city's population (from 12 to 6 million); and (ii) the sudden appearance of extra-provincial officials to entice Dongguan-based firms to leave Guangdong altogether (Interview, Planner A, Shenzhen, January 2013). These developments impelled Jiang Ling, a CPC cadre and then Vice Mayor of Dongguan, to issue a full-page clarification of the city's restructuring policies in a local newspaper, *Xinxi Shibao*, on 29 July 2008. Clearly a bid to ensure social stability (and indirectly to assure senior leaders in Beijing that economic instability was not occurring at all), the rare occurrence of a CPC cadre publicly defending the state's policies only accentuated the difficulty – if not impossibility – of attaining simultaneous industrial relocation and upgrading.

The Guangdong government's ambitious post-crisis restructuring approach not only worried local government officials, it raised the concern of national policymakers. Following the announcement of 'Decisions', the-then Premier Wen Jiabao made two separate trips to Guangdong in July and November 2008 in an official bid to 'investigate and research' the economic situation in the province. For a senior CPC leader to visit a location within China twice in one year is rare; to do so twice in five months is of significance. In his July 2008 visit, less than two months after the launch of 'Decisions' in Guangdong, Wen made clear his feelings on how to proceed with managing the financial crisis:

> Guangdong's socio-economy…is confronted with new challenges. The slowdown in the global economy and contraction in external demand has pronounced impacts on export-oriented medium to small enterprises in the PRD. The proportion of non-state linked medium and small enterprises is more developed in Guangdong. Currently these enterprises are faced with more difficulties, hence support must be enhanced through credit, taxation and industrial policies. (*China Securities Journal*, 21 July 2008; author's translation)

During his second visit and, interestingly, in the company of Wang Yang, Wen Jiabao issued a public reminder that 'the problems and difficulties facing medium and small enterprises have not been fundamentally resolved, their production and management situations remain grave. On this, there should be high-level focus and a continuation of initiatives to help these enterprises overcome their difficulties'. Wen's proposed response, while appearing conservative and paternalistic, was actually predicated on a major principle of governance in China. Contrary to popular conceptions that the CPC prizes GDP growth above all else, any government official in charge of economic development would first be evaluated on his/her ability to maintain social stability – i.e. social relations conducive to perpetual CPC rule – before GDP results are evaluated (cf. the reference to Four Cardinal Principles in Chapter 1). Senior policymakers would not accept the delivery of economic growth on the back of social unrest. It was precisely this fear of social unrest that motivated Wen, the country's top-ranking official in charge of economic development, to adopt a position that prioritised social stability.

Assessed in retrospect, Wen Jiabao's comments in November 2008 exemplify the cross-scalar tensions involved in geographically-specific reconfigurations in China. It was most likely a targeted response to Wang's elaborate view on the issue just a day before Wen's arrival in Guangdong. Made during his visit to the city of Zhanjiang, Wang Yang was defending the Guangdong government's approach to manage medium and small enterprises within the policy framework of 'double relocation':

> Some people say this year alone [2008] 50,000 firms shut down, whether this figure is true is one matter, everyone should seriously analyze this: amongst those firms that have shut down, how many of them are large-scale firms? None! In my prediction, these [fallen] firms were mostly lagging productive forces. The cyclical extirpation of lagging productive forces is an effect of market economic forces.

Reinforcing a distinctive trend throughout his tenure in Guangdong, Wang (once again) peppered his explanation in Zhanjiang in relation to 'market economic forces', *as if* the market has objective rules of its own that the state must follow:

> 30 years ago, we chose the market economy and consequently enjoyed the joy of fast-paced growth. Today we must also courageously confront the pain brought about by cyclical turbulences in the market. Since the Asian financial crisis in 1997, Guangdong has adopted a particular way to run at high-speed, to slow down a little now, to adjust the style and raise the technical capacities suitable for long distance running, this is very normal...the state should never do what the market will not allow, it should never save lagging productive forces. (Ibid.)

Wang's comments were contradictory in two ways, however. First, he reinforced the fact that marketisation in post-Mao China was never about the 'free market' or its supposed corollary, perfect competition. By his own admission, the enterprises hurt most by the 'double relocation' policy were medium and small enterprises; 'none' of the large-scale firms were hurt or, presumably, were allowed to be hurt. Viewed in itself, the survival of large-scale enterprises (of which it was unclear how many were state-owned) is not surprising, as their production processes were (and remain) largely organised through relatively flexible sub-contractual arrangements with smaller (and largely privately owned) enterprises. Of greater significance is the Guangdong government's willingness to accept the subservient and vulnerable position of medium and small enterprises – the economic actors that were supposed to be the centrepieces of a 'free market'. This complicates the popular portrayal of Wang as a market-oriented reformer and illustrates the decidedly protectionist undercurrent of the restructuring strategies (ref. discussion of protectionism in Chapter 2). In fact, the intention was not only to ensure the big businesses were shielded from the crisis; it was to provide these businesses with new opportunities for capital accumulation in other parts of the province (or, failing which, in other parts of the country).

Second, the sequence of development presented in Wang's comment in Zhanjiang foregrounds how it was the CPC that *chose* market-oriented governance, not the other way round. The overarching 'national strategy', this chapter argues, was to enable the CPC to remain in perpetual power (ref. the previously mentioned Four Cardinal Principles in Chapter 1). At the policy level, this means having to experiment with new ways of engaging market regulatory logics without compromising the structural coherence of the party–state. After the 2008 global financial crisis, the Chinese party–state apparatus again found itself having to choose how to strategically reposition economic geographies across the country. Operating within a context of intensifying global interconnections, however, meant the remaking of state–market relations through state rescaling could not be based on a fixed playbook of objective market rules – the process was proactive, opportunistic and experimental at the same time (ref. Figure 3.1, Chapter 3).

And the inherent risks involved in this multi-dimensional regulatory approach explain why the 'nationally strategic' designation of reform frontiers invites 'strategic disobedience' (ref. Wedeman 2001; Chapter 2). Indeed, the Guangdong government's decision to 'never save lagging productive forces' and persist in reconfiguring the provincial industrial structure was just as risk-laden as Wen Jiabao's imploration to keep afloat medium and small enterprises through new interventions. Senior policy-makers thus had to take a chance on whether the tentative approach in Guangdong would reproduce – if not enhance – national economic growth. Just days following Wen's November 2008 visit, Wang offered an oblique counterpoint that reiterated his decision to take this risk:

> Amongst these debates, some are for, some against. I thought about it, over the three decades of 'opening up', Guangdong has embarked on its own path, so let others do the debating. Right now it is still the same, we are taking our own path, a path of scientific development, so let others do the debating. Regardless of what others say, the 'double shifts' must be emphasized, 'emptying the cage and changing the birds' must be emphasized. We must never launch just about anything in order to guarantee GDP growth.

Wang's insistence was the final straw for some senior members of the Politburo. On Christmas day of 2008, the *People's Daily* published a critical commentary – rarely levelled against senior CPC cadres – to express their views on Wang's restructuring plan:

> Just a while ago, medium and small enterprises encountered external and internal difficulties and were effectively immobilized. This especially applied to

labor-intensive medium and small enterprises labeled as 'two highs, one low' [high pollution, high energy usage and low efficiency], they were viewed as impediments to industrial upgrading. Some places have appeared too hasty in the process of 'emptying the cage and changing the birds', this caused a significantly squeeze in the survival space for medium and small enterprises. In the face of the strong force brought about by the global financial crisis, the situation of these enterprises has only exacerbated. (*People's Daily*, 25 December 2008a)

While not mentioning Wang by name, the intended audience of the message was explicitly stated through the references to Wang's 'emptying the cage to change the birds' metaphor and the industrial classification term of 'two highs, one low' (*lianggao yidi*):

Even if there are indeed medium and small enterprises that embody the 'two highs, one low', there should not be a simplistic and brutal squeeze of their survival space. Be it in the enhancement of industrial upgrading or in the relocation of these enterprises, policy and financial support must be given. (Ibid.)

Following this public opprobrium, Wang's tone softened. 'Emptying the cage and changing the birds does not mean emptying out all the birds', Wang explained, 'even a fool will not do this'. By implication, the Guangdong restructuring approach was never a 'foolish' one-size-fits-all strategy. Wang further reiterated the stance taken earlier by the Dongguan government:

Even to labour intensive industries and lower-order manufacturing sectors, the 'empty the cage and change the birds' method cannot be based on compulsion. Right now no enterprise is forcibly made to move, no enterprise is having its water or electricity supplies cut, our method is to attract through benefits, we tell enterprises the places where operating costs are low, the places where workers are aplenty, in practice we respect the will of the enterprises, the government uses appropriate policies and hope people make these kinds of choices [i.e. relocate].

Yet field research indicates that the Guangdong government's 'hope' was officially transposed into a performance target: local officials' ability to successfully implement the 'double relocation' policy could affect their promotional prospects (Interview, Planner C, Shenzhen, January 2013). Interestingly, the formal inclusion of this performance target indicates not all officials were supportive of the 'double relocation' policy. After all, as Planner C explains, local village collectives have been highly dependent on rental collection from migrant workers in cities such as Dongguan, Jiangmen and Foshan, key manufacturing hubs in the PRD. To push for manufacturing firms and migrant workers to relocate could mean a loss in short-term income if not a local economic contraction, a situation from which Dongguan has yet to fully recover (Interview, Planner C, Shenzhen, January 2013). In other words, a developmental path had been established in these manufacturing hubs that benefited specific interest groups such as village collectives, and naturally these groups resisted change that would compromise their economic interests. Changing paths meant local officials had to seriously consider turning this regulatory 'hope' into reality.

It would be useful, at this point, to situate Wang's restructuring attempt within the broader context of state evolution in China. As the preceding discussion has indicated, while the CPC nationalised means of production during the Mao era, the regulation of socioeconomic life was largely left to cadres in individual communes (Chapter 2). Deng and his successors would subsequently allow local governments to propose initiatives for reforms (e.g. the nascent attempt to institute the Household Responsibility System in Anhui, the bottom–up suggestions to implement SEZs in Guangdong, etc.). Through generating this new political logic, the CPC implicitly created room for the previously mentioned 'strategic disobedience'.

In China's long and entrenched hierarchical chain of command, Wedemen (2001: 71) argues, 'structurally induced ambiguities create opportunities [for cadres] to engage in wilful disobedience because they imply that wilful disobedience will go unpunished'. In the case of Guangdong, the entire economy was in a state of uncertainty during global economic slowdown; yet this uncertainty provided an opportune platform for the provincial government to launch its 'double relocation' spatial project. As then-Vice Governor Governor Zhu Xiaodan acknowledged, 'even without the global financial crisis, it was already time for industrial transformation and upgrading in Guangdong' (*Nanfang Dushibao*, 8 March 2012b). In other words, 'strategic disobedience' was in the pipeline independent of the global financial crisis; the new ambiguities generated by the crisis only catalysed the implementation of 'double relocation'. Wang put this point into sharper perspective:

> Regarding extra-local relocation, if it occurred during the good times enterprises would be unwilling to move, yet if it occurred during economic contraction some enterprises claim to have no ability to move, then no movement will ever occur. Practically the financial crisis has lowered the costs of relocation, adding to this the government is providing a series of preferential policies to facilitate industrial transformation and upgrading, this moment is an excellent opportunity for [industrial] relocation. (*Xinxi Shibao*, 18 July 2009a; author's translation)

Placed in relation to the post-2007 reconfiguration of regulatory relations in Guangdong, however, Wang's initiative was more accurately a form of strategic *alignment* with central goals. The Guangdong government's apparent 'disobedience' was not a lack of 'compliance' to an a priori directive from the central government. While the strongly worded response in *People's Daily* may be read as a mandate from Beijing to the Guangdong government, the issuance of this directive was actually a response to a bottom–up developmental approach.

The cause–effect relation in this process is the direct opposite of Wedeman's (2001) assumption that 'strategic disobedience' occurs as a reaction to an a priori directive by the central government.

In addition, there were strong signs the central government had different views on the restructuring process in Guangdong. Under conditions of ambiguity, as then-Premier Wen Jiabao demonstrated in his response to the global financial crisis, the Chinese central government would most likely be cautious in its response. This corresponds with Wedemen's proposition. However, as will be shown shortly, Hu Jintao's subsequent endorsement of the Guangdong project indicates not all policymakers within the central government believed in a cautious response to ambiguities. 'Disobedience' from lower hierarchical levels could thus be symptomatic of differences within the top echelons. Whether its unrelenting stance vis-à-vis calls to help medium and small enterprises would morph into a form of 'disobedience' – and hence become subject to what the CPC calls 'party discipline' – was contingent on its ability to capitalise on planning uncertainties within the central government.

More accurately, then, the Guangdong government's apparent 'strategic disobedience' expresses the dynamic politics of launching experimental developmental projects in China (ref. Chapter 3). It demonstrates how a proposed change in regulatory logics at one scale (Guangdong) inevitably triggers a response from another (the central government). This phenomenon corresponds with Peck's (2002: 340) observation that 'the present scalar location of a given regulatory process is neither natural nor inevitable, but instead reflects an outcome of past political conflicts and compromises'. Yet, as the next section will explain, 'strategic disobedience' did not occur for its own sake. Along with the 'double relocation' project, Wang simultaneously launched the 'scaling up' of territories deemed 'strategic' for the reforms of national institutions. What was initially a provincial-level concern (economic restructuring) was turned into a national issue in order to gain central governmental support – this is a defining characteristic of what the next section terms 'decentralisation as centralisation'.

The Politics of Producing 'Nationally Strategic' Socioeconomic Spaces in Guangdong

As it soon became clear, Wang's overt unwillingness to follow Wen's advice led to speculations over differences in the central government's view of Guangdong's 'double relocation' approach. When then Chinese president Hu Jintao visited Guangdong in March 2009, Wang said in Hu's presence that 'last year [2008], during which our province was confronted by the global financial crisis and was faced with its most difficult moment since the turn of the century, the General Secretary [i.e. Hu] entrusted comrade [Li] Changchun to conduct an investigation of Guangdong and, based on his report, clearly issued opinions that guide our work in Guangdong' (*Yangcheng Wanbao*, 7 March 2009). No mention was given to Wen's two trips to Guangdong in 2008; Wen's aforementioned instructions to save medium and small enterprises affected by the global demand slump also did not figure in Wang's speech. This omission is significant in three ways.

First, the most literal point was that Wen Jiabao did not visit Guangdong with Hu Jintao's instructions; Hu only 'entrusted' Li Changchun, another senior cadre in the Politburo, to visit Guangdong by the Hu government. Second, Wang was indirectly saying Wen's comments had no impact on his or Hu's view on Guangdong's policies – only Li Changchun's report had any impact. And this leads to the third point: the central government's eventual guidance for Guangdong did not include Wen's advice. That Wang could take this daring discursive step strongly suggests he had the full support of the Chinese president in the central government. In turn, it suggests the central government was unsure of its strategic response to the global financial crisis. To be sure, where political polarisation existed within the central state apparatus, it did not paralyse the proposed restructuring project in Guangdong. If anything, as Hu put it in his second visit to Guangdong in 2009, it propelled the project in a forward motion:

> In confronting the global financial crisis, the Guangdong government emphasized seizing the favorable circumstance induced by the crisis to reconfigure and enhance the industrial structure and transform the mode of economic development...I think your reasoning is very good, you have to persevere, always move ahead, and truly give a good fight in this tough battle of transforming the economic development approach. To gain awareness early, to act early, is good proactivity. (*Guangzhou Daily*, 30 December 2009).

Hu's affirmation did not mean Wang's restructuring strategy was objectively correct; what it did was to mark the central government's final position along the spectrum of opinions on China's post-crisis development. With this affirmation, any speculation of uncertainty within the central government was quelled. Full-fledged restructuring, as indicated by economic–geographical transformations within Guangdong, was identified as the way forward.

At one level, Hu's decision could be taken as a victory for bottom–up governance in China: the Guangdong government successfully pushed through its restructuring agenda and 'won'. Perhaps more significantly, the crisis offered an opportune moment for the Guangdong government to create new 'scaling up' opportunities as part of the restructuring project: while the reported large-scale shutdown of enterprises may constitute an excruciating short-term blow to employment and income, the Guangdong policymakers simultaneously sought to ease the economic pain by imploring the central government to demarcate Hengqin, Qianhai and Nansha as 'nationally strategic'. This active engagement with the central government complicates simple models that view centralisation and decentralisation as a binary (ref. review in Chapter 2). As the case studies show in this book, local governments seek not to create autonomous 'feudal economies' (*jingji zhuhou* 经济诸侯) – there are immense political and economic incentives to ensure their local developmental agendas are aligned to, if not at the forefront of, macro developmental goals set by the central government. In this regard, state rescaling becomes a function of a political system that prioritises local policy experimentation in its quest for structural coherence.

Field research reveals Wang Yang had a direct role to play in the national designation of Hengqin and Qianhai. As a Shenzhen-based planner explains, Wang targeted and was intent on setting Guangdong's two SEZs – Zhuhai and Shenzhen – on new developmental trajectories:

> When [the Guangdong Party Secretary] Wang Yang first went to Zhuhai in April 2008, he claimed 'Zhuhai must open up a new path that is different from others in the PRD', the next year [in April 2009] he claimed Zhuhai had enhanced its positioning on the back of its past developments and the future is to be desired. A few months later the Hengqin development project was announced.

Similarly, the spatial relationality of Qianhai only changed after Wang Yang identified it as a strategic location for cross-scalar interaction:

> The Shenzhen authorities have long viewed Qianhai as a good location, but it was only in August 2008, when Wang Yang specially came to inspect the area around Qianhai and Shekou [the original site of Deng Xiaoping's pioneering reforms in Shenzhen] that he pinpointed Qianhai to be the principal fulcrum of Shenzhen's institutional innovation. Thereon he called for a deeper consideration of how Guangdong, Hong Kong and Shenzhen could develop cooperation avenues. Because of this Qianhai's development finally appeared on the planning agenda, it also finally brought into practice cooperative planning, which people have paid lip service to for years, between Guangdong and Hong Kong. (Ibid.)

Through successfully pushing for the 'scaling up' of Hengqin and Qianhai (and later Nansha), it could be argued that the Guangdong government had adroitly coalesced two complementary spatial projects in order to ensure the province retains its leading economic position within the economic planning hierarchy without experiencing either structural unemployment or GDP decline (see next section).

The situation becomes nuanced when assessed from a cross-scalar perspective, however. As Peck (2002: 338) has observed, 'socioregulatory processes operate across scales, rather than being confined to a particular scale, highlighting the need to consider the relative power of scale-based rule systems vis-a-vis that of scale-bound actors, agents, and institutions, a relationship that is often an asymmetrical if not a hierarchical one'. Hu's embrace of radical change could not be reduced to an unproblematic approval of policies proposed by the 'scale bound' Guangdong government; indeed, he arguably embraced change only because new economic–geographical conditions to retain capital – which in turn supports employment, the key source of social stability – within Chinese state spatiality have become possible.[8] It is through this approach that the Guangdong government generated 'relative power' vis-à-vis the central government. This phenomenon corresponds with Peck's (2002: 340) observation that:

> Even if the prevailing pattern of change in the contemporary rescaling process is national-local, this does not mean that the national state qua scalar actor ceases to play a significant role. Rather, its role is reconceived and restructured, in part to engage in more active processes of scale management and coordination at the local and international levels.

If anything, the ability of the central government to grant provincial and municipal governments the authority to launch new policy innovations exemplifies its 'reconceived and restructured' role. This arguably contains echoes of a Mao-era mode of spatial organisation that treats local governments as 'cellular' units (cf. ref. discussion in Chapters 1 and 2). As one experienced urban planner in China remarked over a lunch conversation, 'whenever the central government launches a project, many people at the local level automatically respond with much fervour. There is no problem getting people [at the local scale] to do things' (Personal communication, Beijing, February 2012; author's translation). This comment corresponds with those of Zhou Xiaochuan (of the People's Bank of China) and Lu Dadao (of the Chinese Academy of Sciences), as presented in Chapters 1 and 3.

Viewed against this broader backdrop of intense inter-governmental competition across contemporary China, this book argues that the 'double relocation' strategy was a means to achieve rescaling. The pain of relocating undesirable industries in Guangdong was gradually eased because the 'scaling up' of Hengqin, Qianhai and Nansha as 'nationally strategic new areas' guaranteed a significant amount of fixed capital investments. This in turn kept GDP figures positive – as shown in Figure 4.2 – in spite of the financial crisis. To this end, Wang had to enrol the economic and extra-economic institutions of the central state apparatus into the 'double relocation' process. On the one hand, Wang needed the central government's political support to rescale the economic–geographical configuration in Guangdong. On the other hand, the proposed designation of targeted territories required the financial support of key central government-owned enterprises through loans or direct investments.

Ironically, as a series of interviews in Shenzhen and Macau revealed, while Wang Yang was espousing the virtues of 'the market', the construction projects in Hengqin and Qianhai were made possible by 'special' loan approvals from the Chinese Development Bank and the large-scale participation of central

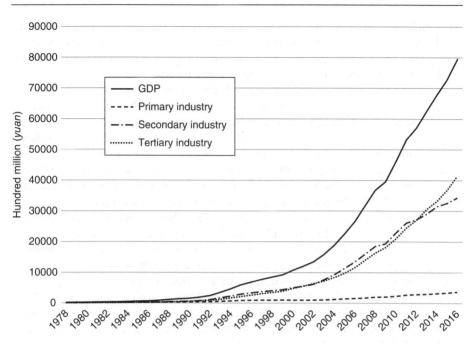

Figure 4.2   GDP growth (total and sectoral) in Guangdong, 1978–2016. Source: Based on data from Statistics Bureau of Guangdong (2013).

government-owned enterprises (yangqi 央企). In addition, locally-based SOEs were allocated land use rights to parts of Hengqin and Qianhai, sensing state rescaling could lead to new profit-making opportunities, these SOEs were unwilling to relinquish their land-use rights to planning bureaus. This problem was particularly acute in Qianhai, where planners faced delays in actualising their plans due to the conflicting interests between these different state-linked groups. Widely-stereotyped as an embodiment of the CPC's embrace of deepening marketisation, the reforms launched by the Wang administration more accurately reflects how subnational agendas impact the structural coherence of the national political economy (ref. Chapter 2). The CPC, in short, is never independent of the market.

The financial impact of the 'nationally strategic' designation was felt almost immediately. Specifically, the rescaling process jumpstarted fixed capital formation in the three cities where Hengqin, Qianhai and Nansha were located, namely Zhuhai, Shenzhen and Guangzhou. For Guangzhou and Zhuhai, fixed capital formation in 2012 more than doubled the pre-crisis, pre-new area level in 2006, while the growth in Shenzhen was just under 100% over the same period (Guangdong Bureau of Statistics 2013). These growth figures for Guangzhou, Shenzhen and Zhuhai was unusually high, considering the three cities were already highly built up. The only exceptions were the areas marked as 'nationally strategic', all of which were empty land plots at the onset of the financial crisis. Working from these statistics, the research proceeded to investigate whether short-term short term fixed capital formation in these three new areas contributed strongly to fixed capital formation of these three cities.

Fixed capital formation exceeded 33 billion *yuan* (~US$5.45 billion) in Hengqin New Area between January 2010 and December 2012, the end of Wang's tenure as Guangdong Party Secretary (*Nanfang Ribao*, 7 November 2013). The figure increased to a total of 114.7 billion yuan (~US$16.95 billion) at last count in February 2016 (*Zhuhai Tequbao*, 11 March 2016). Measured against the total fixed capital formation statistics for Zhuhai between 2010 and 2012 (in which Hengqin is located), fixed capital formation in Hengqin contributed almost 20% of the Zhuhai total; by the end of 2015, the proportion increased to 22.2% (*Zhuhai Tequbao*, 11 March 2016; Zhuhai Statistics Information Network; author's calculations). Hengqin New Area has therefore emerged as a major contributor of fixed capital investments in Zhuhai (and, by extension, Guangdong province), a development unforeseen prior to Wang's tenure. While figures for Qianhai were not announced, a budgetary plan drawn up by the Qianhai Bureau and the China Development Bank revealed that between 2013 and 2022, 480 billion yuan (~US$79.2 billion) would be budgeted for fixed capital formation in the rescaled plot of 15 km². Of this sum, 266.5 billion yuan (~US$43.9 billion) would go towards basic infrastructural construction (*Jingji Guanchabao*, 3 August 2012). Quite clearly, then, state rescaling generates a positive economic outcome for local governments in China. Before new regulatory capacities were even introduced in these reconfigured territories, local GDP figures would have inflated because of the large-scale fixed capital investments.

By extension, the growth in fixed capital formation associated with economic-geographical 'scaling up' produced a trend that contrasted those of most crisis-hit city-regions in East Asia. According to official figures revealed by the Governor of Guangdong, Zhu Xiaodan, more than 7000 firms were relocated (though it was unclear whether they relocated within the province or moved to other regions within China) between 2008 and 2012, while a staggering total of almost 80,000 firms either stopped operations or shut down during the same period. Intriguingly, against these signs of economic contraction, GDP and employment growth in Guangdong remained positive through the post-crisis period (i.e. 2009 to the time of writing).

On the one hand, the figures on firm relocation reinforced what former Party Secretary Wang Yang – the chief architect of the restructuring program – said in relation to potential short-term shifts in restructuring: 'the state will not save lagging productive forces'. On the other hand, it justified Wen's concern that economic restructuring would significantly affect the livelihoods of medium and small enterprises. Yet statistical calculations from the Guangdong Statistical Bureau painted a beautiful picture: the GDP for all major economic sectors grew. More importantly, contrary to Wen's public announcement that

200 million people in China was unemployed in 2010 alone, there was no reported decline in the unemployment of registered residents in the Pearl River Delta (no mention was given to the number of migrant workers 'relocated' from the city-region, interestingly). Something had to be done to ensure GDP growth was not compromised – and Wang arguably did it through pushing for selected territories to be designated 'nationally strategic'.

To understand why the provincial-level income and employment figures remained positive, it is instructive to begin with the announcement that the function of internal demand in GDP-generation grew from 80.4% in 2007 to 89% in 2012 (*WenWeiPo*, 26 January 2013). This effectively meant, in proportional terms, the total reliance on external demand to generate GDP halved during this period. In the same period, 30,000 new enterprises were brought in, generating investments worth 108 times those of firms that relocated or shut down (Ibid.). Read in relation to the 270% growth in fixed capital formation for the Pearl River Delta city-region between 2006 and 2013, it can be further ascertained that the Guangdong government's short-term recovery strategy in the immediate post-crisis period was to ramp up fixed capital formation in what was already a highly developed region (Guangdong Statistical Bureau 2014). After all, as Niu Jing, Director of Hengqin New Area Administrative Committee, acknowledges, 'we need to build a lot of buildings for commercial, residential, hotels and other facilities in the coming years' (*South China Morning Post*, 2 September 2013).

Crucially, it was because of the potential GDP contribution that policymakers in Guangdong were willing – and able – to take on the financial and regulatory risks associated with short-term, large-scale fixed capital investments in rescaled areas of Hengqin, Qianhai and Nansha.[9] To finance investments in fixed capital, these governments committed to taking on loans or issue bonds. The Hengqin New Area Administrative Committee, for instance, announced it had plans to issue 1.5 billion yuan (~US$250 million) in bonds in Hong Kong and was planning to list a portfolio of property and infrastructure projects on the city's stock exchange in 2014 for about 2 billion yuan (~US$310 million) in proceeds (Ibid.). This financing strategy came on top of the 'specially approved' loans it had obtained from the China Development Bank to finance earlier rounds of land requisition (Interviews, Macau and Shenzhen, January 2013).[10] Even before business transactions are conducted in the individual new areas, the rescaling process already contributed significantly to the provincial GDP. This underscores the central conceptual point: through engaging the central government in its experimental projects, the provincial government could continue to generate economic growth. Clearly the central government cannot engage all provincial or municipal governments at once in its experimentation, which in turn opens up room for intense inter-territorial competition (the 'competitive alignment' as discussed in Chapter 3; ref. also comments by Lu Dadao and Xiao Jincheng in Chapter 1). It is in this respect that regulatory relations across scales get re-negotiated in contemporary China.

Conclusion

That economic development in Guangdong had been driven by labour-intensive and highly pollutive industries – particularly since the intensification of urban-oriented reforms in the mid-1990s – is well documented. That this trajectory could not last was also expected. Quite how the province was to undertake structural reforms without undermining its economic importance at the national scale was unknown, however, until the 'double relocation' industrial policy became defined by 2009. With the provincial economy increasingly impacted by the 2008 global financial crisis (and which, at the time, showed no signs of abating), the Wang administration made a drastic announcement to categorise redundant economic actors as 'lagging productive forces'. A form of managed deindustrialisation ensued. Viewed in tandem with the Guangdong government's high-profile push to 'scale up' the intra-urban territories of Hengqin, Qianhai and Nansha into 'nationally strategic' reform frontiers (ref. Figure 4.1), the 'double relocation' strategy may not appear to be directly of the 'national interest'. On closer inspection, however, this industrial policy was a means to enable the institution of new reforms and policy experimentation in Guangdong. In other words, the emergence of new regulatory logics in Guangdong entailed a shift in its underlying industrial structure, a prerequisite that would not be met without the proactive intervention of the provincial policy-makers. This chapter has provided a critical evaluation of this intervention.

Two conceptual implications can be drawn from the state rescaling process in and through Guangdong. On the one hand, Hu's decision reflects the growing importance of bottom–up governance: the Guangdong government successfully pushed through its restructuring agenda and 'won'. Perhaps more significantly, the crisis offered an opportune moment for the Guangdong policy-makers to create new 'nationally strategic' opportunities as part of the restructuring project: while the large-scale shutdown of medium and small enterprises may constitute an excruciating short-term blow to employment and income, the Guangdong policy-makers simultaneously sought to ease the economic pain by imploring the central government to designate the territories of Hengqin, Qianhai and Nansha as 'nationally strategic' reform frontiers (ref. Figure 4.2 on positive GDP growth between 2008 and 2012). The Guangdong Free Trade Zone (GFTZ) was then developed following the establishment of these three frontiers. This active engagement with the central government complicates simple models that view centralisation and decentralisation as a binary, as previously discussed in Chapter 2. It calls, instead, for a systematic evaluation of industrial restructuring in the GPRD – multiple policies in different locations were instituted simultaneously between 2008 and 2012, each engaging with different inherited institutions, each generating new relations with the central government and transnational capital.

On the other hand, the persistence of a fragmented national economy is not engendering a resurgence of autonomous 'feudal economies'. Rather, it demonstrates how the system of reciprocal accountability, to re-borrow Shirk's (1993) terms, has generated immense political and economic incentives for the *competitive alignment* of local developmental agendas to the centrally determined 'national strategy' (ref. Chapters 2 and 3). This in turn calls for the relationship between Wang Yang and senior policy-makers like Hu Jintao and Wen Jiabao to be evaluated at the macro-structural level. Despite their differences, all three were embedded within the system of reciprocal accountability and were expected to behave strategically within the logics of this system. Wang's strategic objective was to ensure the PRD city-region (and Guangdong more broadly) remain attractive to capital. His practical goal was to assure both the central government and corporate investors that the province remains an attractive site of/for capital accumulation. Against the context of falling global demand, this emphasis was sufficient in itself to convince Hu and Wen that the potential economic gains outweighed concerns about social instability.

Placed within a broader geographical–historical context, the 'double relocation' policy in Guangdong indicates state rescaling in China is not an end-state that can be attained or secured. As discussed in Chapter 3, state rescaling is not a mechanical, one-way devolution of regulatory capacities from the national scale to the supra- or subnational governments. To demonstrate this, this chapter moved beyond an uncritical mapping of new state spaces to explain how the reconfiguration of regulatory scales functions as an ongoing process of contestation, negotiation and co-management of Chinese state spatiality. In so doing, it foregrounded the contradictions associated with national-scale regulatory directives, some launched back in the 1950s (ref. Peck and Zhang 2013; Zhang and Peck 2016). Specifically, the post-crisis restructuring agenda in Guangdong reaffirmed the difficulty – if not impossibility – of attaining Mao and Deng's joint vision of national spatial egalitarianism. The central government's willingness to approve this agenda, which has since been emulated by other coastal city-regions in Zhejiang and Jiangsu, further undermines attempts by the Jiang Zemin and Hu Jintao administrations to institute more 'coordinated' regional development (ref. Chen and Ku 2014). These contradictions collectively underscore the difficulty of actualising Pareto-optimality in inter-regional resource transfers within the context of intensifying global economic integration. Indeed, insofar as place-specific competitiveness in the global economy remains a core policy concern for the CPC, it appears that new rounds of policy experimentation and reforms in China will continue to be shaped by the tensions between state rescaling initiatives, such as the path-generating agendas introduced in Hengqin, Qianhai and Nansha New Areas, and the tendency, if not temptation, to traipse along national-level pathways instituted by earlier regimes. The next chapter will examine the emergence and effects of these tensions.

Endnotes

1   The GDPs of Hong Kong and Macau are calculated separately from mainland China.
2   See 'Dedicated Chapter' on Hong Kong and Macau, Special Column No. 22.
3   The urban-bias of export-based economic development is reflected in the economic–geographical disparities between the PRD and the rest of Guangdong (see Lu and Wei 2007; Sun and Fan 2008). To legitimise the 'double relocation' program, the Guangdong government argued that relocating less desirable industries to poorer regions could reduce these disparities.
4   Officially, given the lack of inter-provincial coordination, it has to be stated that relocation policies remain within the province so that the government appears to prioritise the provincial interests. This is an extension of the protectionist, inward-looking tendencies of the Mao era. In reality, however, given the huge difference in infrastructural facilities between the PRD and the rest of Guangdong province, the Guangdong government in fact did not mind if undesirable industries leave the province altogether rather than attempt to upgrade these industries within the province (Interview, academic and regular consultant to the government, January 2013).
5   The official term of this project in Mandarin is <朱江三角洲地区改革发展规划纲要>. The cited content is printed on page 7 of the original document; translated by author.
6   This term that would subsequently be widely used to characterise Guangdong's province-wide restructuring approach.
7   Hu's support was important at both the personal level and in terms of national economic–geographical reconfiguration. In 2007, when Wang Yang was Party Secretary of Chongqing, Hu approved a series of nationally-strategic socioeconomic policies aimed at bridging the urban–rural divide caused by the Mao-era household-registration (or hukou) institution of population control. That Hu would approve these major reforms during Wang Yang's term was a glowing recognition of the latter's efforts at initiating institutional reforms. Ironically, it was precisely the institution Wang tried to reform in Chongqing – the *hukou* system of demographic control – that enabled the Guangdong government to effect labour force relocation at will (ref. Chapter 2 for an overview of the socioeconomic regulatory logics of this institution; Wang's role in driving the *hukou* reforms in Chongqing will be discussed in Chapters 6 and 7). As introduced in Chapter 1, this phenomenon at once exemplifies the fragility and the flexibility of the growth-based social contract in the quest to accumulate capital in China.
8   Hu approved in 2010 the Chongqing government's proposal to demarcate Liangjiang New Area as a 'nationally strategic new area'. Put forth by Wang's successor, Bo Xilai, and the-then Chongqing mayor, Huang Qifan, the high profile establishment of Liangjiang New Area was developed in tandem with the large-scale social reforms planned during the Wang Yang leadership. For this reason, Chongqing became a major location for industries looking for more cost-effective locations within China (the Chongqing reforms will be explored in Chapters 6 and 7). Viewed in relation to each other, it is clear that 'nationally strategic' reforms in Chongqing and Guangdong were not as antagonistic as portrayed in the popular media. If anything, their roles are complementary.

9   What is intriguing in this instance is the strong willingness of SOEs (including banks) owned by the central government (the *yangqi*) to support fixed capital formation in the rescaled spaces. With risks undertaken by the provincial and SEZ governments in Guangdong, the involvement of these SOEs effectively allows the central government to profit from the construction of these rescaled 'nationally strategic' spaces. There thus exists a strong economic incentive for the Chinese central government to partake in state rescaling – provided, of course, the local governments do not end up defaulting on the loans they undertook to launch fixed capital investments.

10   The involvement of the China Development Bank in the rescaling process arguably deserves separate research focus. Preliminary verifications of these interview claims showed that (i) an initial sum of 200 million yuan was advanced to Qianhai in 2012 for infrastructural construction (*21st Century Business Herald*, 3 August 2012); (ii) an undisclosed sum was loaned to the Hengqin planning committee to compensate those affected by land requisition, the costs of which were also shared by the Macau government (Interview, Macau, January 2013); while an agreement was signed between CDB and Nansha New Area for a loan of 60 billion yuan (*Yangcheng Wanbao*, 13 November 2013). Rescaling, quite clearly, involves large-scale and recurring financing, and the Chinese central government is in a commanding position to offer this financial support after it consolidated its power through the 1994 fiscal reforms and the deepening of SOE restructuring in 1998. The transference of 'move first, experiment first' regulatory capacities to local scales thus further reaffirms central state power by opening up new opportunities for economic control – this is the defining characteristic of what is conceptualised as 'decentralisation as centralisation' in this off (ref. Chapter 8).

Chapter Five
Becoming 'More Special than Special' II
Hengqin and Qianhai New Areas as National Frontiers of Financial Reforms

---

Introduction

On 25 July 2012, the Guangdong provincial government released for public feedback the draft 'Decision on several problems pertaining to the total advancement of financial reforms to enhance provincial development' (hereafter 'draft Decision').[1] Building on an agenda approved in June 2012 by the Chinese State Council, this announcement offered a more detailed definition of new economic–geographical developments that built on the 'double relocation' program in the Pearl River Delta (PRD) (ref. Chapter 4). First, it reflects a targeted attempt at reconfiguring national state space. The goal is to address 'nationally-strategic' constraints associated with the financial structure, some of which were introduced in Chapter 2. While the 'draft Decision' broadly defines Hengqin, Qianhai and Nansha New Areas as platforms for financial innovation, each territory is to accomplish strategic functions. Policy experimentation in the creation of an off-shore capital market would be launched in Hengqin; the goal was to allow investment trust funds to be established in multiple currencies, in turn serving as a platform for onward private equity financing on the Chinese mainland. Building on this model, the Hengqin administrative bureau subsequently collaborated with Qianhai and Nansha New Areas to form the Guangdong Free Trade Zone (GFTZ). Reflecting the intense inter-locality competition within the Chinese

On Shifting Foundations: State Rescaling, Policy Experimentation and Economic Restructuring in Post-1949 China,
First Edition. Kean Fan Lim.

political economy, the professed goal of the GFTZ is to emulate China's first FTZ in Pudong New Area in Shanghai (*Tencent Finance*, 10 May 2015).

Connecting to and re-constituting Hong Kong's RMB [renminbi, the official currency of the People's Republic of China] credit market is the designated role for Qianhai. The key experimental objective is to test out the financialisation of domestic investment projects through offshore RMB trading centres (of which Hong Kong is currently the most developed). If experiments succeed in Qianhai, the Communist Party of China (CPC) would have produced a new economic–geographical tool. Specifically, new onshore, non-financial 'backflow mechanisms' would be established for RMB money markets located in 'offshore RMB trading, with the CPC having total leverage on the geographies of these 'backflow' investments (ref. Section 6.2). The experiments in Nansha are more tightly linked to the actually-existing 'material economy' (or *shiti jingji* 实体经济): it is to experiment with policies that would enhance the functioning of existing and new industries. The primary policy challenge for Nansha, which mirrors the challenge encountered by the entire province, is the degree to which private banks and/or credit houses could engage in the financing of new investments by medium and small enterprises (collectively referred as *zhongxiao qiye* 中小企业 in China).

By 'scaling up' Hengqin and Qianhai New Areas, the Wang administration involved the central government and the two Special Administrative Region (SAR) governments[2] in negotiations, with the aim of obtaining the right to experiment with policies that would (once again) place Guangdong at the forefront of national reforms. Primarily this involved obtaining the much-desired 'move first, experiment first' power from the central government. As discussed in Chapter 3, experimental measures not subject to censure could be launched in so far as they were deemed to attain objectives set by the central government. Yet there was no a priori guarantee the Guangdong government could succeed: it had to propose potential institutions for reform that the central government would consider to be of national significance. To this end, the Guangdong government competitively aligned its developmental agenda with the dilemmas confronted by Zhou Xiaochuan, governor of the People's Bank of China (China's central bank) and a key proponent of financial reforms for many years (Interview, Planner C, Shenzhen, January 2013). While keen to push through reforms, Zhou (2012: n.p.) had earlier expressed worries over the tension between integrated, 'chessboard'-styled planning and bottom–up initiatives:

> Because the financial market is based on dynamic flows, regardless of whether reforms are targeted at specific functions or localities, the potential for spillovers is very strong. The original intention of reforms may be to restrict experiments to a particular domain, but the practical effects are often uncontrollable, which means there will be positive or negative effects on related domains or adjacent regions. Restricting reformist experiments to specific domains may even result in unfair competition. Which means to say, the financial sector was supposed to be regulated

systemically, on the basis of 'the country as one chessboard', but in the event that some places enjoy specific reformist policies and some places do not, or they enjoy other types of reformist policies, this may result in the unfair allocation of financial resources. In short, the spillover effect may be quite big and hard to control, and may affect the monitoring and evaluation of reform effects.

Read in relation to the 'draft decisions', it becomes clear that the designated areas of Hengqin and Qianhai were geographically-delimited to preclude 'spillovers' from policy experimentation. In the Hengqin case, a re-bordering exercise was introduced, while regulations were introduced to circumscribe cross-border credit for investments to Qianhai-based firms. Through this territorial reconfiguration, 'the monitoring and evaluation of reform effects' would be less problematic. More importantly, locating the reform sites in Guangdong would build on the proximity to Hong Kong and Macau, both 'free' economies that are well integrated with the global system of capitalism. For Zhang Bei, the current Head of the Qianhai Bureau, the unique interaction between the national, local and the global scales (through Hong Kong) was the raison d'être of Qianhai New Area, with geographical advantage distinguishing it from competition elsewhere in China:

> From an external perspective, first the RMB internationalisation progression is accelerating, which means the window for financial innovation in Qianhai is narrowing; second, different locations are intensifying competition for business capital, they similarly position themselves as sites for modern services, especially finance-related services, leading to homogenising competition; third, policies are also becoming increasingly similar, in some domains, the support given to some policies are even not as good as those in the mainland. (Interview with *Nanfang Dushibao*, 15 August 2013b; author's translation)

Similarly, Hou Yongzhi of the Development Research Centre (DRC), a highly influential policy think tank in the Chinese State Council, believes the 'one country, two systems' institution involving Macau and Hong Kong was pivotal in the demarcation of Hengqin as a site of financial reforms:

> Within the country there are…quite a number of places try to launch the concept of 'new areas', Hengqin stands out in two ways, namely the backdrop of the 'one country, two systems' institution and the fact that it is based on cross-border regional collaboration. Hence its model can be summarised in two phrases, one, it is only through collaboration that the development of this 'new area' could be advanced, and two, it is through the development of this 'new area' that win-win frontiers will open up. (Transcribed by *Nanfang Daily*, 24 August 2012; author's translation)

This cross-border connection was arguably why the 'move first, experiment first' policies were granted to Hengqin and Qianhai ahead of Shanghai, a major municipality directly under the control of the central government. The proximity would allow the central government to launch and monitor two major reforms, namely capital account convertibility and RMB internationalisation.[3] In short, the Guangdong government won the 'race' for preferential policies because of its ability to benefit from and reshape actually-existing economic geographies. Yet this 'race' was only the beginning of the rescaling process; the longer-term challenge was whether the newly-designated zones could drive economic growth within their broader city-regional contexts. This chapter will assess and evaluate the cross-border flows that undergird this challenge.

Divided in four parts, the chapter will first consider in the following section the primary external factor that contributed to the designation of 'nationally strategic new areas' in the PRD – the establishment and function of Hong Kong as an official offshore RMB centre. The third section examines how the re-bordering of Hengqin facilitate economic integration with Hong Kong and Macau. Taking a similar approach, the fourth section examines the implications of Qianhai's 'nationally strategic' role. The two sections will both evaluate whether the new policies will allow the CPC to overcome one major constraint – freedom of capital flows – that confronts its regulation of the national financial structure. The concluding section will then delineate the conceptual contributions of the case analyses.

Hong Kong's Emergent Functions as an 'Offshore RMB Center'

The 'offshore RMB center' is an emergent economic–geographical contradiction. It is at once a spatial expression of the CPC's desire for financial liberalisation and a conduit for the CPC to extend its heavy-handed control of financial flows. The geographies of 'free' RMB circulation and usage outside Chinese state spatiality must be officially 'designated' by the CPC. Hong Kong is the first and, with around 50% of the world's offshore RMB deposits in 2014 and accounting for 70% of global RMB-related financial activity, the biggest designated RMB offshore trading centre in the world at the time of writing (Hong Kong Legislative Council 2015: n.p.). Many financial centres around the world were and continue to be vying for the CPC's approval to operate new RMB offshore centres. The political decision not to adopt capital account convertibility, an approach long viewed as a disadvantage in international trade, has offered significant geographical leverage for the CPC: it could now determine the flows and fixity of the RMB at different scales (ref. Box 5.1).

The reason why Hong Kong plays the role as the leading offshore RMB centre is geographical as well as historical. Since the 1978 'liberalisation' reforms, Hong Kong has been a targeted source of capital as well as the 'jump off' point for exports. Its proximity to the SEZs enabled the city to take on new hub-like functions in finance and logistics. Since the 1990s, following deepening reforms of SOEs, Hong Kong became the city of choice for many restructured SOEs

---

Box 5.1   Is Capital Account Convertibility (Un)Necessary for Capital Accumulation?

Standard economic textbooks define capital account convertibility as the freedom to convert local financial assets into foreign financial assets and vice versa. Conversion prices are to be determined by the market, without any form of state intermediation or regulation. There are typically three components within the capital account (in some countries also called 'Financial Account'), namely 'Foreign Direct Investment', which encompasses investments in a local incorporated entity or a joint venture with a local entity, and which are deemed unlikely to pull out at short notice; 'Portfolio Investment', which covers transactions in stocks and bonds that do not influence the day-to-day operations of domestic firms (i.e. investments that are less than a predetermined percentage of shares deemed sufficient to give decision-making rights); and 'Reserve Assets', which are assets (including foreign exchange reserves) that can only be bought and sold by monetary authorities such as central banks (e.g. the People's Bank of China). The 'Capital Account', like all other components of national income accounting, presupposes the existence of the nation-state: they exemplify and help reinforce state-centrism.

One corollary of capital account convertibility of a particular national economy is the ability of anyone, regardless of citizenship, to freely move the currency of that particular economy across its geopolitical borders without any form of regulatory intervention or barrier. Another corollary is anyone could invest in any asset priced in the currency of the particular economy with a fully-convertible capital account. There are very few economies that adopt this ideal-typical model; even the United States, a long-time proponent of fully-convertible capital accounts, mandates anyone who enters the US with more than US\$10,000 to make declarations at border checkpoints. Foreign investments in US-based assets are routinely evaluated in relation to 'national security' requirements. Contrary to popular wisdom, there is currently no universal institution that defines and enforces capital account convertibility. The lack of universal adaptation of the ideal-typical model capital account convertibility thus makes it ironic that the model is widely portrayed in the mainstream media as the sine qua non of 'efficient' participation in the global economy. It should more accurately be depicted as a political project to facilitate financial market integration, under the (unproven) assumption that it represents a more 'efficient' allocation of financial capital in world markets (cf. Minsky 1986; Harvey 2010).

Interestingly, debates on the necessity of full capital account convertibility have become fierce amongst economists (the very inventors of this model). In an empirical analysis provocatively titled 'Who Needs Capital Account Convertibility?', Rodrik (1998) argues that no correlation exists between the openness of countries' capital accounts and investment amounts or their respective growth rates. Because of this lack of correlation, it is difficult to ascertain the benefits of an open capital account. Conversely, Rodrik (1998) argues, the costs of open capital accounts are regularly expressed through crises in emerging markets. In the same year, Levine and Zorvos (1998a, b) published two studies that demonstrate how capital account liberalisation had no effect on investments. The Asian financial crisis – which occurred just as these papers were published – lent further empirical credence to their claims. This important finding extends to the theorisation of the Chinese economic growth 'miracle', especially in the past decade, within which GDP growth was primarily attributed to large-scale, state-led investments and a selective opening of the capital account (i.e. the foreign direct investment [FDI] component, although it remains subject to regulatory approval and sectoral barriers to entry). It appears quite straightforward, then, to affirm that the ideal-typical model that presents a positive correlative relationship between capital account convertibility and economic growth would not stand up to empirical testing in China.

But the picture gets complicated when one takes into account Hong Kong and Macau, two economies that come close to fully adopting the ideal-typical model, are extra-territorial jurisdictions *of* China. The roles of these economies (particularly Hong Kong, though Macau has increasingly become important after it superseded Las Vegas as the largest gambling capital of the world) as conduits of capital flows to/for mainland China are well-known. The more interesting development, as the Hengqin and Qianhai cases show, is the economic–geographical attempt to support the functions of Hong Kong and Macau. Does this attempt suggest the CPC is willing to extend the 'universalisation' of the capital account convertibility model to the entire state territory, or is it, in its classic Maoist approach to development, an affirmation and negation of this model at the same time?

Source: Author's compilation

---

seeking equity financing. In 2008, the-then Chinese Premier, Wen Jiabao, designated Hong Kong and Macau as the first centres for cross-border, RMB-denominated trade settlements. The scheme was later extended to encompass the Association of Southeast Asian Nations (ASEAN) economies, and was officially launched in July 2009.[4] The primary reason why these centres were chosen was to enable Chinese companies to reduce their exposure to the exchange risks of US dollar (cf. Lim 2010). This unprecedented policy, as Eddie Yue, Deputy Director of the Hong Kong Monetary Authority, explains in 2012, injected a fresh dimension to Hong Kong's role as a hub connecting China and the global economy:

> Given the pace of change, it is sobering to recall that it wasn't until 2009 – not all that long ago – that the Mainland authorities started to allow Mainland corporations to

use the RMB when trading with the rest of the world…Over this period of liberalisation, a rapidly growing offshore RMB market has also emerged, centred in Hong Kong. The rising level of RMB trade flows to and from the Mainland has produced an expanding pool of RMB liquidity. And the growing liquidity in the offshore RMB market has led to the development of RMB financing and forex markets, with an estimated daily turnover of around 2 to 4 billion US dollars for spot and forward transactions. (Eddie Yue, Speech on 23 May 2012: n.p.)

With this new development, Yue continues, Hong Kong's relations with the global financial system received a boost:

More foreign banks are either choosing to have a base in Hong Kong or are using RMB correspondent banking services provided by banks in Hong Kong. As of the end of March 2012, there were close to 200 banks participating in Hong Kong's RMB clearing platform. Of these, 170 are foreign-owned or located overseas. At the same time, over 1100 RMB correspondent accounts were maintained by overseas banks with banks in Hong Kong. The amount due to and due from such overseas banks amounted to RMB 128 billion yuan and RMB 146 billion yuan respectively, a clear indication that banks from around the world are using the robust platform and large liquidity pool to offer RMB services to their customers at home. (Eddie Yue, Speech on 23 May 2012: n.p.)

At the core of the RMB settlement processes in Hong Kong is a new market-based institution for the RMB. Known as the 'offshore RMB market' or CNH, it allowed the RMB to be traded at market prices. The rate in the mainland, dubbed CNY, continues to be subject to constant intervention by the People's Bank of China. Two geographically-differentiated exchange rates thereby coexist for the RMB, with the rate in Hong Kong always lower than that on the mainland. As a result, new opportunities for exchange rate arbitrage emerged. Since the RMB is under constant pressure to appreciate, a distinct CNH–CNY spread currently exists. Because less RMB is needed to buy the same amount of US dollars in Hong Kong than in the mainland, a now-common response from importers in the mainland would be to purchase US dollars in Hong Kong to pay for their imports. As an interviewee explains (Interview, Planner B, Shenzhen, January 2013), a typical strategy of arbitrage involves the 'round-tripping' of commodities (which correspondingly also leads to the inflation of trade figures discussed in Chapter 2):

- Step 1: A Chinese merchant borrows US$1 million from a bank based in mainland China. S/he then changes the funds at an onshore rate of US$1: RMB6.25, buying RMB6.25 million.
- Step 2: The Chinese merchant's offshore company in Hong Kong buys a specific amount of gold valued at RMB6.25 million. This gold is then re-sold to the merchant's mainland Chinese firm at RMB6.25 million, for processing

in one of Shenzhen's three 'bonded ports'. No import taxes are incurred on the transaction.

- Step 3: The merchant's HK company instantly changes the RMB6.25 million 'revenue' back into US dollars at the lower offshore rate of US$1: RMB6.1. This buys US$1.024590 million. The RMB6.25 million becomes 'offshore RMB deposits' in the HK capital market.
- Step 4: In Shenzhen, the gold is re-processed into simple jewellery and re-exported to the merchant's HK company at a valuation of US$1 million. The deal is settled in US dollars, which sees US$1 million flow from HK to China. From this re-exporting process, the mainland company makes US$24,590 (US$1.024590 million–US$1 million). The HK company then re-sells the processed jewellery to retailers at higher than US$1 million, or holds it until gold prices rise higher than the original US$1 million value before melting and selling the gold. More profits could potentially be made on the HK side.

As Figures 5.1 and 5.2 show, the concentration of RMB deposits and authorised RMB business operators surged after the 2009 cross-border trade announcement. At the time, there was an added incentive for foreign businesses to hold the RMB – the widely-held sentiment that the RMB was undervalued and would eventually be revalued or be increasingly priced by market forces (ref. Lim 2010).

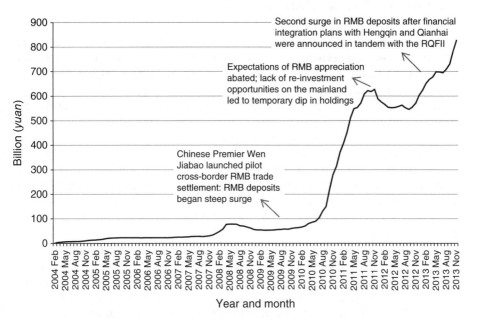

Figure 5.1   Total RMB deposits in Hong Kong, 2004–2013. Source: Hong Kong Monetary Authority. Author's graphical and textual illustration.

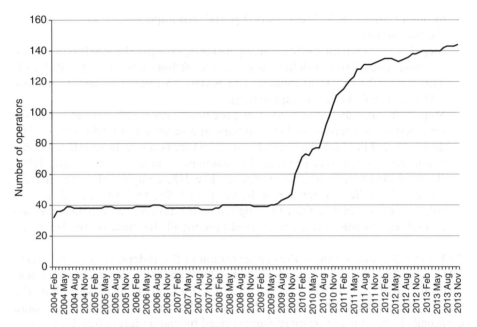

Figure 5.2   Number of authorised RMB business operators in Hong Kong, 2004–2013. Source: Hong Kong Monetary Authority. Author's graphical and textual illustration.

When expectations of sharp revaluation did not become reality (the Chinese state continued to resist international pressures to 'free float' the RMB), RMB holdings in Hong Kong dropped between November 2011 and August 2012. This underscores Hyman Minsky's (1986: 228) observation that 'everyone can create money; the problem is to get it accepted'. That 'problem' was soon eased after the Hengqin and Qianhai financial reforms planned were unveiled in July 2012. In what appears to be a highly-coordinated move, the Hong Kong Monetary Authority announced on the same day as the Hengqin/Qianhai unveiling that non-residents in Hong Kong were allowed to purchase unlimited RMB at the offshore rate (*Bloomberg*, 25 July 2012). Given the new Hengqin and Qianhai policies present unprecedented re-investment channels in the Chinese economy for investors based offshore, a second surge in RMB holdings in Hong Kong ensued (ref. Figure 5.1).

Given the rapid increase in RMB concentration in Hong Kong at the time, new reinvestment mechanisms for offshore RMB deposits in Hong Kong must be established in order to retain interest in using the RMB as a means of exchange. Data published in late 2013 reveals RMB-denominated cross-border settlements between Chinese mainland firms and Hong Kong clearing institutions to be only 10% of the 400 billion (~US$66 billion) in daily RMB settlement, with 90% being 'purely offshore' (*China Daily*, 18 October 2013). This means RMB settlements predominantly circulate either between Hong Kong and overseas financial markets or amongst overseas markets via Hong Kong.

There are two paradoxical reasons for this phenomenon. First, foreign companies and banks were (and are still) willing to transact in the RMB, in anticipation that its exchange value will increase vis-à-vis that of the dollar (the value of which has depreciated steadily over the last decade). Second, RMB flows outside the mainland financial system are 'trapped' as they could not find adequate investment opportunities in the Chinese economy. In David Harvey's (1982) parlance, there are few 'spatial fixes' for this circulatory capital. To encourage more merchants and institutions to conduct RMB-denominated transactions, more onshore–offshore 'bridging' mechanisms had to be introduced in order to direct some of these 'purely offshore' transactions into the mainland Chinese economy.

Prior to new channels introduced in Hengqin and Qianhai, the primary 'bridging mechanism' in a designated RMB offshore trading centre was the granting of licences to companies that qualify as a 'Renminbi Qualified Foreign Institutional Investor' (RQFII). RQFII-licensed firms could channel RMB funds raised in the offshore trading centre (e.g. Hong Kong and London) into securities markets within mainland China. At the onset of its launch in December 2011, the quota for reinvestment was set at a relatively low 20 billion *yuan* (~US$3.3 billion). Subsequent demand saw the quota revised multiple times; in January 2017, the quota for RQFII funds stood at 530 billion *yuan* (~US$77.2 billion).

Specifically, RQFII funds must invest primarily in RMB-denominated bonds, bond funds and/or shares issued in mainland China. Offshore RQFII holders (typically multinational banks like HSBC and Standard Chartered) may issue public or private funds or other investment products using their allocated RQFII quotas. Non-RQFII activities in the offshore RMB centres are geographically-delimited to those centres, e.g. RMB-denominated bonds issued within Hong Kong or London (also known as 'dim sum bonds' for their relatively small size relative to bonds issued within mainland China), the ability to open RMB bank deposit accounts, etc. Non-RQFII financial products are significantly less attractive because they cannot be re-invested into non-financial sectors within the Chinese economy. Owing to appreciation expectations and the new opportunities for investment in RMB-denominated assets in the mainland through the RQFII policy, RMB deposits and bond issuance in Hong Kong have increased correspondingly.

The logic that undergirds the RQFII initiative is straightforward: it allows the CPC to determine RMB 'backflow' channels in the first instance. And it was arguably this initiative that set the precedent for the Qianhai policies. The ability to control 'backflow' mechanisms would enable monetary policy to be implemented more effectively even as the RMB is 'internationalised'. This contrasts the approach of governments adopting a fully-convertible currency, in which the currency could technically remain in offshore capital markets and be immune to the objectives of domestic monetary policies (the US dollar and Japanese yen are two clear examples of this offshore–onshore disconnect). Eddie Yue of the HKMA acknowledges the fundamental role these 'backflow' mechanisms play in the functions of an offshore RMB market:

> [T]he offshore RMB market is a market of a currency that is not yet fully convertible, and the links with the onshore financial markets are still subject to limitations. The RMB is convertible in respect of the current account and certain capital account items such as direct investments, while other flows like portfolio investments can only be conducted under specified quotas. The opening up of even more channels for two-way flows between the onshore and offshore markets would be required to spur offshore RMB activities. (Eddie Yue, speech on 8 May 2013)

As Yue continues, this development is the outcome of the CPC's willingness to break out of its self-instituted path of capital controls:

> A few years ago, these ideas would have sounded difficult, but today we are edging ever closer to realising many of them, which is indicative of the rapid pace of change in the offshore RMB market. But as we all recognise, this will be a paced process and not something that would or should happen overnight. On this front, the HKMA has been engaging with Mainland Chinese authorities on further expanding fund flows between the onshore and offshore RMB markets. (Eddie Yue, speech on 8 May 2013)

Viewed in relation to the HKMA's push for more 'two-way flows between the onshore and offshore markets', this book argues that the Hengqin and Qianhai financial reforms constitute spatial strategies to 'further expand fund flows between the onshore and offshore RMB markets'. The implementation of these strategies strongly suggest the backflow mechanism offered by the RQFII – which is limited only to investments in securities markets in China – is in itself insufficient. As mentioned earlier in this section, the primary reason is economic–geographical: the RQFII scheme does not allow the CPC direct influence over where funds are to be invested within Chinese state space. In other words, the CPC is trying to broaden the scope of backflow mechanisms that offer more flexibility to respond to uneven conditions across the concrete (non-financial) economy. If the Hengqin and Qianhai experiments were successful, the CPC would obtain new economic–geographical leverage: it could direct RMB-denominated portfolio investments to targeted locations within China, with the associated risks underwritten by investors based offshore.

At the macro-level, then, RMB offshore centres function as an economic–geographical paradox: the 'freedom' enjoyed by institutional investors to invest in China is a function of the CPC's control of the Chinese financial system. This contradiction challenges one widespread speculation, namely that these centres exemplify the Chinese central government's inevitable embrace of full capital account convertibility (ref. Box 5.1). It might be more accurate, rather, to conceptualise these new paradoxical expressions as the experimental 'offshoring' of centralised financial control. Shortly after these policies were introduced, Chan Yan Chong, an analyst of Chinese financial reforms at the City University of Hong Kong, argued that the Chinese central government was not about to release the brakes on capital account convertibility any time soon:

> So far, at least 31 mainland cities have vowed to be China's next financial centre. However, to this day Hong Kong can still only operate a very limited yuan business because the central government has a very cautious attitude towards the convertibility of the yuan. The Binhai New Area in Tianjin vowed to build a world-class financial centre by allowing mainland investors to directly invest in Hong Kong's stock market, but it eventually fell into disuse because Beijing didn't want to make a hasty decision on the convertibility of the yuan. (Chan, interview with *South China Morning Post*, 5 September 2012)

As the next two sections will argue, the 'nationally strategic' designations of Hengqin and Qianhai must be assessed in relation to (i) Hong Kong's evolution as an offshore RMB centre designated by the Chinese central government; and (ii) Macau's growing importance as a global hub of capital concentration after its gambling industry became the biggest in the world (by revenue) since 2007. As early as March 2008, a reporter from the British Broadcasting Association (BBC) Chinese service noticed Wang Yang, then newly-appointed Guangdong Party Secretary, actively engaging the Chief Executives of Hong Kong and Macau at the annual national meeting in Beijing. The situation was uncommon given the three leaders had just met two months before. Not long thereafter, Wang pronounced the future of the Greater Pearl River Delta (GPRD) should be reconfigured 'from the perspective of the world' (*BBC*, 22 April 2008).

It is now increasingly clear that the state rescaling through Henrqin and Qianhai is defined by the simultaneous realignment of central control and local developmental agendas. For the central government, the production of these territories is part of a broader experimental attempt to produce 'backflow channels' for the RMB through the 'perspectives' of Hong Kong and Macau. Capitalising on the national concern to institute 'backflow channels' by locating the first of these channels just across the Hong Kong and Macau borders with the mainland, Wang was not only able to gain a 'first mover' advantage over other provinces; the transfer of 'move first, experiment first' power to Hengqin and Qianhai arguably cushioned the economic blow generated by the 'double relocation' restructuring policy through bringing in new fixed capital investments and guarantees of further RMB inflows from Hong Kong and Macau (ref. Chapter 4). Whether liberalisation ultimately occurs at the local level will not be under the purview of the provincial government – it has been 'scaled up' to a national issue.

At the theoretical level, the production of Hengqin and Qianhai raises a broader question pertaining to macroeconomic governance in China: is full capital account convertibility really the end goal of financial reforms in China? Or could capital account convertibility be averted – and hence allow the CPC to retain full price-setting power on the domestic RMB exchange rate – because new 'backflow' spatial fixes could be found? As Box 5.1 indicates, there is no scholarly consensus that full capital account convertibility confers an automatic competitive advantage in the current global economy. Financial experimentation in Hengqin and Qianhai could thus further challenge predictions that financial reforms in China will produce a historical inevitability – a fully convertible RMB.

At Once Within and Without: The 'Extra Territorialisation' of Hengqin New Area and its Role in Cross-border Financial Integration

Following the approval of the development plan for Hengqin in 2009, the island quickly evolved into a free trade zone and a huge capital market. On 23 May 2012, the Zhuhai government promulgated the 'Administrative Measures on Business Registration in Hengqin New Area of the Zhuhai Special Economic Zone' (SEZ) (hereafter 'Measures'). As reflected in the 'Measures', the threshold for business registration was lowered and the registration procedures simplified. The entwinement of a 'liberalised' registration system and a 'stringent' supervision system, in the official parlance, was intended to establish a solid foundation for the creation of an international business environment in Hengqin New Area. Yet it was clear right from the beginning that this 'liberalised' environment could not exist if Hengqin remained integral to Chinese state territory 'proper'. To change economic life, the CPC first changed space.

The centrepiece of the production of Hengqin New Area was arguably its extra-territorialisation. From the 2009 plan, it was clear that the most striking policy innovation lies in the formation of special immigration and customs clearance rules involving Hengqin, Zhuhai SEZ and Macau SAR. An unprecedented phenomenon in the history of 'new China', the objective was to create a new regulatory space in economic actors from the Chinese mainland could freely enter and engage in production and trade without going through immigration. Only goods going in and out of Hengqin New Area from the mainland side are subject to checks and taxes; there is no stipulation of the amount of RMB these actors could bring into Hengqin, a development which will be central to financial reforms (more on this shortly).

The overall plan for Hengqin New Area involved deepening integration with Macau. A new 'second line' customs post to inspect the flow of goods to and from Hengqin was completed by 2014. The location of the post, just by the Hengqin Bridge, is adjacent to the new Shizimen Business District, strategically located within Hengqin's new borders to capitalise on and effect the special policy provisions on the island. This customs post will clear goods to be imported into and exported out of mainland China proper. Immigration checks will be loosened at what is officially termed the 'first line' border between Hengqin and Macau. As Table 5.1 shows, a trial scheme of localised rules on imports, taxation and immigration was launched in Hengqin on 1 August 2013. Under the new regulations, machinery and equipment imported for use in Hengqin infrastructure and in manufacturing plants will not be taxed. Consumer goods and construction materials imported from Macau will not be exempt from tax. Companies located on the mainland side will still have to pay import and sales tax on products that are manufactured and processed on Hengqin.

The experimentation with new trade and currency flows through Hengqin illustrates the same caution exercised by Deng Xiaoping in 1978: the engagement with transnational capital flows are to be spatialised in targeted territories rather than enforced nationwide 'in one cut' (*yidaoqie* 一刀切). This caution raises two interrelated questions on contemporary Chinese socioeconomic reforms. First, does the formation of 'free' economic spaces like Hengqin indicate further concessions by the CPC to the pressures of transnational capital, or are these spaces reflective of the CPC's attempt to influence the global flows of capital through place-specific experimentation? Second, does the willingness to experiment with currency convertibility inside Hengqin signify an advanced step towards full capital account convertibility at the national scale, or does it reflect a nascent tendency to institute capital account convertibility only in places the CPC deems suitable for its macroeconomic regulation?

While it is not possible to arrive at definitive answers in this nascent period of reforms, specific signs and contradictions are emerging. After the newly-promulgated customs regulations, equity fund investment guidelines and policies on currency flows are compared, it is clear that the role of Hengqin is to 'extend' the deep and well-developed capital market in Hong Kong into the mainland. The new financial regulations stipulate that Fund Enterprises can invest in non-listed companies, private placement stocks issued by listed companies and engage in other relevant services in accordance with the relevant State-level regulations governing portfolio investments by equity funds. This offers a new avenue of investment for Hong Kong-based institutional investors, particularly those without the permission to invest directly in the Chinese financial markets (ref. Section 5.2).

At one level, these new provisions are similar to those promulgated in the 'new areas' of Binhai (in Tianjin) and Pudong (in Shanghai). Two features stand out, however. First, special provisions are given to financial institutions based in geographically proximate Hong Kong and Macau. Working in tandem with newly-promulgated labour regulations to allow professionals from Hong Kong and Macau to be treated as 'extra territorial' personnel in Hengqin (i.e. they will be subject to the personal income tax rate of Hong Kong and Macau instead of those in mainland China), equity funds and fund management companies in Hengqin can directly access the much developed labour market in financial services in Hong Kong (and on a smaller scale, Macau).

Table 5.1   Key aspects and implications of newly promulgated 'On Measures to Supervise and Administer Hengqin New Area by the Customs of the People's Republic of China (Trial Run)', launched 1 August 2013.

Aspects	Corresponding articles in 'Measures'	Implications
Two-tiered cross-border movements of goods and people	• **Article 2:** The customs shall subject these measures to the supervision and checks of transportation equipment, cargo, articles moving in and out of Hengqin, as well as the supervision and checks of enterprises and places in Hengqin that are registered with the Hengqin customs. • **Article 3:** 'First line' regulation is established between the Hengqin and Macau customs; 'Second line' regulation is established between Hengqin and other places within the People's Republic of China (hereafter 'external places'). The customs shall implement differentiated regulation on the principle of 'first line loosening, second line control, human-cargo segregation, and regulation by categorisation'.	• Unprecedented cross-border arrangement for a territory previously located within the Chinese mainland • Underscores the importance of 'free' flows of commodities and capital (expressed in the flow of more affluent and more skilled visitors from HK and Macau) in an increasingly open economy
Geographical determination of commodity inflows	• **Article 9:** Apart from the following goods, the customs shall manage all other goods are tax-free or tax-protected: (i) Consumables and goods used for private real estate construction; (ii) Goods that are not legally allowed to be tax-free or tax-protected; (iii) The list of goods deemed not to be tax-free or tax-protected by the Ministry of Finance, the Tax Bureau, the Customs and related departments • **Article 10:** Apart from those stated in the law, administrative legislation and other regulations, no importation quotas or approval documentations are needed for cross-border movement of goods into Hengqin. Cross-border movements of goods out of Hengqin are subject to export quotas and approval documentations.	• Subjects foreign banks to the mandated geographies of investments/loans • Interestingly, the flow of finance – which is technically a commodity – is not restricted between mainland China 'proper' and Hengqin; this omission explicitly allows Hengqin to serve as a platform for RMB 'internationalisation'
Key regulatory emphases	• **Article 31:** All measurements (termed 'customs statistics') are confined to the movement of physical goods; no reference made to movement of financial capital	• Facilitate integration of production networks, particularly with firms registered in Macau and Hong Kong

Source: Articles from General Administration of Customs of China (GACC); author's translation and analysis.

Second, firms investing in Hengqin's financial services could enjoy a flat corporate income tax of only 15%, an unparalleled comparative advantage vis-à-vis two of the mature 'nationally strategic new areas', namely Pudong and Binhai. With these policies, the Chinese government is, in the words of an interviewee in

Macau, literally 'serving up China's private equity (PE) market at the door steps of Hong Kong and Macau' (Interview, Academic B, January 2013, Macau). Yet it is important to note that the creation of PE hubs do not translate into capital account convertibility. All venture capital (in RMB or foreign currencies) must still be subject to monitoring by the officially termed 'custodian banks'. Relative to Qianhai, which functions narrowly as a new credit market for institutions based in Hong Kong and Macau (more on this shortly), the experimental policies in Hengqin facilitated the creation of a new capital market.

Crucial to this new capital market, however, is the continued restrictions of capital circulation to and from Macau and Hong Kong. It could thus be argued that the financial reforms in Hengqin are not novel. Rather than institute unregulated currency convertibility, experiments with 'free' convertibility are at present circumscribed to the opening of bank and credit card accounts in either the Macau pataca or Hong Kong dollar. Furthermore, the Hengqin authorities must first approve banks that could issue these 'foreign' currencies. A Shenzhen-based planner who participated in discussions on the Hengqin plans sums up the current stage of uncertainty:

> If you look at the official document, the positioning of Hengqin's development states that 'on the premise of cooperation, innovation and services, Hengqin's locational advantage to connect Guangdong with Hong Kong and Macau should be maximised; this should enhance tighter cooperation and integrated development with Hong Kong and Macau, gradually develop Hengqin as a first-mover in liberalisation reforms and scientific innovation, and enable Hengqin to become a new model for developing the "one country, two systems" institution.' You can see that these points are all macro overviews, or, in other words, vague. (Interview, Planner B, Shenzhen, January 2013)

Planner B went on to elaborate on a specific area of uncertainty:

> Particularly unclear is the point about 'one country two systems', what are the forbidden zones of this 'new model' in the modification of the 'one country, two systems' institution? Be it at the level of Guangdong province, or Shenzhen, Zhuhai and other cities that are linked to Hong Kong and Macau, what are the areas that are allowed to achieve breakthroughs in reform? I think even the central government doesn't have the answers. (Interview, Planner B, Shenzhen, January 2013)

Placed in relation to the offshore RMB institutions in Hong Kong, it is unclear what function Hengqin would play to deepen financial integration in the GPRD. Going by the view of Niu Jing, Director of Hengqin New Area Administrative Committee, the successful engagement of Hong Kong appears to be highly significant in the Committee's plan: '[w]hile Macau is our closest neighbour, Hong Kong is our partner' (*South China Morning Post*, 2 September 2013). An analysis of the experimental policies indicates, however, that Hengqin is not in a position to function as an independent 'offshore' RMB centre like Hong Kong. The existence of a porous 'second line' border means it could not function as a delimited 'spatial absorber' of offshore RMB backflows like Qianhai (see next section). Indeed, despite the re-bordering guidelines introduced in Table 5.1, RMB flows into Hengqin could 'leak' back to the mainland relatively easily through the 'second line' border (Article 31 indicates no controls on domestic currency flows in and out of Hengqin are instituted). Indeed, as one Macanese legal consultant points out, Hong Kong was first an 'offshore' financial centre before it formally took on RMB businesses, whereas Hengqin was never institutionalised as a unique offshore economy like Hong Kong:

> Actually one important reason why Hong Kong became the official offshore RMB accounting centre a decade ago was because a lot of RMB was already in 'illegal' circulation. And now they are saying Hong Kong's global financial standing will benefit from its ability to push forth 'RMB internationalisation'. But Hengqin has neither the growing offshore RMB supplies nor an existing offshore financial system, so its ability to become an offshore RMB centre is dependent on the scale of off-shore RMB transactions it can accommodate. This will depend on two things, whether the national financial policymakers have such a vision [of Hengqin as an offshore centre that will be eventually like Hong Kong] and whether the major financial institutions are willing to locate there. Without these factors, things will not work out, there will be no independent financial market in Hengqin. (Interview, Macau legal consultant and academic, January 2013; author's translation)

This observation is important in two ways. First, Hong Kong's emergence as an offshore RMB centre in 2003 was not only unintended; it originally was not part of a 'nationally strategy' to facilitate 'RMB internationalisation'. To be sure, the emergent pool of RMB deposits was only 'illegal' relative to the PRC's law to strictly confine the circulation within domestic borders. That this 'illegality' is now taken as a competitive advantage for Hong Kong is not only ironic, it reflects and is an out-come of the contradictions between the CPC's national-level financial regulations and the demands of transnational circulatory capital (ref. Chapter 3).

As discussed in Section 5.2, the growing impetus to reduce transactions through the US dollar entailed the creation of new channels to enhance the cross-border flow of the RMB. The actually-existing 'illegal' concentration of RMB in Hong Kong thus provided a good platform from which to begin: if more investors could freely access the growing concentration of RMB in HK and re-direct these investments into the Chinese mainland, trade settlement through the RMB would increase. Hong Kong-based investors could in turn take on the risks of investing in the Chinese mainland (via Hengqin's emerging PE market) without having to first convert US dollars into RMB. Similarly, if firms based in Hengqin could access offshore RMB funds in Hong Kong through loans or issuing bonds, there would be an added avenue for the CPC to direct the 'backflow' of RMB from offshore locations.

By extension, then, the reforms in Hengqin represent an attempt to test whether deeper integration with offshore capital markets (currently Hong Kong and Macau) could occur without compromising the CPC's need to retain control of the domestic financial system. While these reforms are highly contingent on Hong Kong's role as the designated 'offshore transaction centre' (*li-an jiesuan zhongxin* 离岸结算中心), it is unclear how Hengqin would function as a strategic 'partner' in 'internationalising' the RMB relative to Qianhai. Unless offshore RMB can flow freely into the PE capital market in Hengqin, following which new onward investments – which, as Figure 5.3 shows, functions as the 'leakage' of offshore RMB funds into the mainland – in non-financial sectors could be made, it appears Hengqin's primary function may be to facilitate free trade. If so, it is not a revolutionary phenomenon, given that 'bonded ports' across China perform exactly the same role.

Viewed in relation to regional economic integration, the financial reforms in Hengqin seem to have the strongest impact on the Macau financial sector at the time of writing. As one interviewee states, 'Binhai and Shanghai are at the end of

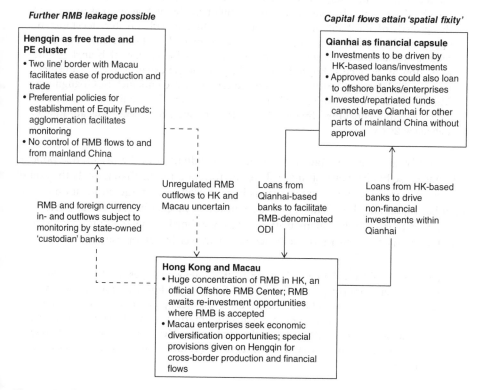

Figure 5.3   State rescaling and financial restructuring in the GPRD: the emergent functions of Hengqin, Qianhai, Macau and Hong Kong. Source: Author's illustration.

the day hubs created by the Chinese government for the Chinese mainland, the Hengqin institutions operate on the basis of extending the long-term prosperity of Hong Kong and Macau and the Chinese government's commitment to diversify Macau's economic structure' (Interview, Academic A, Macau, January 2013). Indeed, following the implementation of the Guangdong-Macau Cooperation Agreement, Macanese banks now enjoy distinct entry preferences in Hengqin. Rather than an act of financial 'liberalisation', this move must be understood in the context of the CPC's commitment to 'preserving the long-term prosperity of Hong Kong and Macau', as (re)institutionalised in the 12th Five Year Plan (2011–2015) and reflected in practice through a quasi-free trade agreement known as the Closer Economic Partnership Arrangement (CEPA). This point is well summed up by Mr. Ip Sio Kai, Deputy General Manager of the Bank of China (Macau branch):

> The value of requested year-end total assets for Macao banks that wish to set up a branch or station a legal representative on Hengqin has been reduced to US$4 billion. This is the only mainland region for which CEPA has cut the access threshold for banks. It is also a new policy tailor-made for Macao so as to support the growth of the banking industry by providing it with greater development opportunities…[Before this agreement] Macao-owned banks can only engage in securities and futures margin deposits in mainland China as a bank branch, so the reduction of the access threshold is very helpful for the expansion of this business as well. With permission to set up consumer finance companies in Guangdong, Macao's financial institutions will access the mainland financial market easier so as to expand consumer finance business. (Ip, interview with Macau Image, February 2013: n.p.)

Specifically, as Ip and a Macau-based legal consultant add, the preference policies for Macanese banks play a similar role to the function in Qianhai, namely the ability for an offshore provider to extend cross-border credit to a targeted location on the mainland. Just like the offshore providers of credit in Qianhai are circumscribed to Hong Kong-based financial institutions, the cross-border flow of credit from Macau into Hengqin applies primarily to Macanese firms setting up operations in Hengqin:

> The existing CEPA framework permits the following types of loans, namely, consumer loans, mortgages, brokerage and financing of business transactions. As a testing ground for financial innovations, those that set up on Hengqin are in a position to experiment with cross-border loans for enterprises based there. This means that any local businesses [from Macau] setting up on Hengqin can continue to rely on their familiar financial service providers and get more cross-border services and support. (Ibid.)

While agreeing the involvement of Macanese banks is positive for the small and previously constrained financial sector in Macau, the Macau legal consultant adds that this involvement does not translate into 'free' capital flows (a precondition of capital account convertibility):

> They lowered the barriers to entry for Macau's banks, their involvement [in Hengqin] is interesting, once they have set up shop they would be able to support foreign investments on the island on the basis that regulation of these investments would be thin, if non-existent. It also allows Macanese banks to grow their RMB reserves. I think how these banks engage in business in Hengqin will shape the Chinese government's thinking on capital account convertibility. But *it's still a very controlled experiment*, you know, any inflow and outflow of RMB capital will only be limited to Hengqin. And if you look around, you will see the big state banks are here too. I don't know how they will engage in the support of investment projects and conduct the free conversion of currencies in Hengqin without simultaneously feeding the information back to authorities on the mainland. (Interview, Macau legal consultant and academic, January 2013, emphases added)

The retention of formal monitoring channels notwithstanding, another practical issue with deregulating capital flows between Macau and Hengqin is the extent to which they would replace existing channels of informal flows. The channels for informal flows of currency in and out of the country have existed since the Mao era. Over the past decade, however, these flows have assumed distinct forms. In Zhuhai, the city within which Hengqin is located, a vast underground foreign exchange network in the Gongbei border zone with Macau has been in operation for more than a decade. Following the relaxation of restrictions for cross-border travel in the late 1990s, informal currency flows in this border zone have grown significantly. Exchanges of a few thousand US dollars now take place over the counter, in full view of the public. Larger sums of exchange (more than US$100k per transaction) would take place behind closed doors; hired runners could deliver the foreign exchange or RMB to hotels in Macau or Zhuhai (personal conversation with exchange vendor, Zhuhai, January 2013).

Unlike currency exchanges in banks, where the exchange of any amount of foreign currency entails an identity card (for Chinese citizens) or a passport (for foreign visitors), these over-the-counter exchanges operate on a no-questions-asked basis. No checks are required and only rudimentary receipts are issued (if at all). There is no upper limit to the amount that could be exchanged, which strongly suggests the vendors have access to a large liquidity pool that is unregulated by the Chinese state apparatus. Or rather, as previously discussed in Chapter 2, the authorities have long known of such unregulated channels of capital flows but choose to overlook it. Borge Bakken, professor of criminology at the University of Hong Kong, puts the situation in perspective: 'If you have the political connections in China, it's quite easy to get away with these things. There are thousands of ways to get money out of the country' (*Global Post*, 19 April 2012).

The existence of this fluid network of cross-border currency flows in and through Zhuhai thus exemplifies the ease in which 'hot money' could flow in and out of China, just like it would in any 'free' economy. The strong onshore demand to convert RMB into foreign exchange is matched by significant demand for these RMB in the 'shadow banking system', which is essentially an unlicensed, 'back alley' banking network that seeks to grant loans to high-risk customers (cf. Tsai 2004):

> There is a lot of demand for credit from small businesses on the mainland side, they pay up to 20% in interest, way higher than the bank rate. But there is also strong demand for loans in the formal banking sector. Because there is now strong demand for the RMB in Macau, if the RMB 'makes a run' across the border, it will bring in foreign exchange, which can buy more RMB on the mainland side. This is one key reason why the underground exchange network is thriving. And all these came before the Hengqin project, actually what is happening in Hengqin is of no interference [to this network], they run separately. (Interview Macau legal consultant and academic, January 2013)

If the Hengqin reforms are not intended to undermine these informal channels of currency flows in Zhuhai, how would they enhance the national financial sector? A highly plausible logic is the lack of an underlying intention to lift capital controls. Indeed, the Zhuhai channel notwithstanding, another major avenue for hot money inflows through fake invoicing of exports has also been prevalent for at least a decade. As is well-documented, the hub of fake invoicing is in the Shenzhen–Hong Kong border region, less than an hour away from Hengqin by fast-boat (cf. Figure 4.1, Chapter 4). As illustrated earlier in this chapter, one strategy of exchange rate arbitrage is to 'export' to a subsidiary in Hong Kong; others simply send lorries of exports across Hong Kong–China border, before having the lorries perform a U-turn and transport the goods illegally back into China. If an 'exporter' performs this 'one day tour' with US\$1 million worth of goods 10 times, this would mean with one batch of goods s/he had exported US\$10 million on paper. The export invoice could then be used as collateral to obtain loans of US\$10 million (or its RMB equivalent) in the Hong Kong capital markets (where interest rates are very low relative to the rates issued by the state-linked banks on the mainland).

In November 2013, China's State Administration of Foreign Exchange (SAFE) announced it uncovered 1076 instances of false reporting by 112 companies, adding up to \$2.5 billion (LeBlanc 2014). The more realistic figures appear to well surpass this official account. According to Global Financial Integrity (GFI), a non-profit institute in Washington DC that specialises in researching the illicit flow of money in developing and emerging economies, the trend of hot money inflows in China since 2006 is positively correlated to the US 'quantitative easing' strategies (i.e. expansion of monetary supply). The primary mechanism to facilitate these flows, as GFI economist Brian LeBlanc (2014: n.p.) explains, is through fraudulent invoicing:

> How big is this hole in China's capital controls? Although no estimates from SAFE [the Chinese Government's regulatory body] were provided, over-reported exports are easily detectable through a comparison of bilateral trade statistics. A comparison of China's trade with Hong Kong shows that an alarming $101 billion of exports simply disappeared at the Hong Kong border in 2012 alone, with an additional $54 billion smuggled in during the first quarter of 2013. The cumulative amount of foreign exchange brought illicitly into China masked as trade payments from Hong Kong since the first quarter of 2006 adds up to an astounding $400 billion. Putting this into perspective, the $101 billion of foreign exchange brought in through illicit exports represents about 40% of the $253 billion of legal net FDI that China received from abroad during the same year. Such large sums of money have the potential to be destabilising, and are most likely being used to fuel further currency and housing speculation within the country.

Interestingly, SAFE did not repudiate the GFI study. Neither did it repudiate a Credit Suisse (2013) study that identified a growing trend in 'alternative lending' vis-à-vis that of the formal banking system. Viewed in relation to arguments that capital account convertibility is necessary for economic growth, the GFI and Credit Suisse research strongly suggests economic growth occurs *because of* capital controls: the resultant 'hot money' inflows drive the concrete economy in China in the shadow banking and real estate sectors and directly influence domestic inflation rates. The only effect of capital account controls is it enables the CPC to determine the exchange value of the RMB while retaining the independence to set interest rates.

Against this backdrop, the transfer of new regulatory capacities to Hengqin would at best be to experiment with new methods to control cross-border financial flows. Without a simultaneous attempt to clamp down on informal flow channels (through fake export invoicing and the vast underground exchange network) and tighten currency controls at the 'second line' border, the policy experiments to relax movements of currencies in and out of Hengqin would always be a step behind – if not totally independent of – 'hot money' flows (ref. Figure 5.3). Then again, given the CPC's unrelenting stance on the fixed exchange rate regime and capital controls, that might just be the intention.

Qianhai New Area as an Onshore 'Spatial Fix' for Offshore RMB Flows

Following the designation of Qianhai as a 'nationally strategic' experimental zone, 50% of the demarcated 15 km² area was allocated for use by the financial industry. Another sizeable land area was demarcated for a 'bonded port', which means goods can come in and out of Qianhai without tariffs, so long as the goods do not make onward movement into the rest of China. The Qianhai Financial Industry Plan identified RMB internationalisation as the top priority for the experimental zone's policy reforms. Twenty-two preferential policies were eventually conferred by the Chinese central government to support the cross-border financial integration with Hong Kong (see Table 5.2). In January 2013, the first cross-border RMB loan scheme was launched – 15 Chinese companies located in Qianhai borrowed a total of two billion yuan from 15 banks based in Hong Kong. At the same time, the central government approved another of Qianhai's policy proposal: certain state-owned financial institutions could now raise RMB in the Hong Kong capital market and establish a 'fund of a fund' (similar to a private equity fund) in Qianhai.

Both projects were aimed at promoting the flow of the RMB back to the mainland from Hong Kong, where a substantial and rapidly growing amount of RMB reserves have accumulated (the specific regulations and implications are listed in Table 5.2). As mentioned earlier, geographical proximity to Hong Kong was listed as the top reason for the national designation of Qianhai. As Wang Jinxia, the spokesperson for the Qianhai Management Authority (QMA), puts it: 'It can be said that independent of Shenzhen-Hong Kong cooperation, Qianhai will no longer be Qianhai, it will no longer have any competitive advantage'. The former QMA chief, Zheng Hongjie, puts this relational positioning in more elaborate perspective:

> The service sector takes up 92% of gross domestic product (GDP) in Hong Kong relative to 55% in Shenzhen. Over the past 30 years Shenzhen and Hong Kong have cooperated to relocate manufacturing industries northwards, right now Qianhai could do similar explorations for modern service industries, we feel that this positioning is accurate. Regarding financial development, RMB internationalisation is now on the agenda and is progressively implemented. Hong Kong is a global financial centre, hence Qianhai can make use of its standing and its strength to experiment with new onshore–offshore services, this has received the support of the 'one bank and three commissions' [the acronym for the four major institutions regulating the national financial sector, namely the People's Bank of China; China Banking Regulatory Commission; China Securities Regulatory Commission; and China Insurance Regulatory Commission]. From the angle of industrial transformation and upgrading in Shenzhen, the provision of financial services is also necessary. (Zheng, interview with *Nanfang Dushibao*, 14 August 2013a; author's translation).

Further research into the financial reform policies reveals a blurred picture beyond the initial boost in fixed capital investments in Qianhai, however (cf. Chapter 4, section four). With reference to Table 5.2, while the Qianhai cross-border loan policies nominally preclude further onward 'leakage' into mainland China (ref. Figure 5.3, Table 5.2), there are at least two possible ways firms could overcome these regulations. First, there is no requirement in the regulations that state suppliers of Qianhai's construction projects need to be located in Qianhai.

Table 5.2   Key aspects and implications of the temporary measures on cross-border RMB loans in Qianhai.

Aspects	Corresponding articles in 'Measures'	Implications
Key actors involved in cross-border loans denominated in Chinese *renminbi*	• **Article 2:** Qianhai cross-border loans in RMB refers to domestic firms that, having fulfilled required conditions, borrow RMB funds from Hong Kong banks that offer RMB-related services. The domestic firms that fulfilled conditions refer to those established in Qianhai and are in operation or directly invested in Qianhai. • **Article 5:** Under the guidance of the People's Bank of China (PBoC), the Shenzhen PBoC branch in downtown Shenzhen will supervise services involved in cross-border RMB loans	• Unprecedented process for domestic PRC firms, although another[a] regulation was promulgated in June 2012 for foreign-owned firms to borrow RMB from offshore centres, irrespective of their geographical operations within China • Opens up new – if still small – financing channels for domestic firms • Increases risk exposure of foreign banks to the credit creation process within China; deepens economic integration • Chinese central government remains in direct control of capital flows through Shenzhen PBoC
Geographical determination of investments	• **Article 6:** The RMB loan account balance is supervised by the Shenzhen PBoC, subject to the developmental situation of RMB-related banking services in Hong Kong; the construction and developmental requirements of Qianhai; and the requirements of macroeconomic regulation. • **Article 7:** The usage of the cross-border RMB loans should be used in the construction and development of Qianhai, on the basis of relevant national policies • **Articles 8 and 9:** Loan durations and interest rates are to be determined freely between the parties involved.	• Subjects foreign banks to the mandated geographies of investments/loans • Enhances central governmental ability to shape uneven economic–geographical development through the regulation of financial flow and fixity • Freedom from domestic interest rate controls entails subservient to a new regulatory barrier: to be approved to do business in Qianhai

(Continued)

Table 5.2   (Continued)

Aspects	Corresponding articles in 'Measures'	Implications
Key regulatory emphases	• **Article 16:** The domestic settlement bank [i.e. Shenzhen PBoC] should check on the usage authenticity of funds derived from cross-border RMB loans • **Article 18:** Domestic firms who have to open RMB accounts in Hong Kong in order to facilitate the cross-border loans must report the opening of such accounts to Shenzhen PBoC within 5 working days	• Primary aim appears to be to prevent Qianhai from becoming a spatial conduit for onward-flows of RMB into the rest of the Chinese economy (ref. Figure 5.3) • Mainland Chinese firms subject to onshore monitoring even if they set up operations or accounts offshore

Source: Shenzhen People's Bank of China Notice 173, 2012 (前海跨境人民幣貸款暫行辦法); author's translation.
[a] The regulation is listed out in <The Notice on the details of direct investments in RMB clearance businesses by foreign merchants> (Notice 165, 2012).

In other words, investments would eventually be expressed physically in Qianhai, but the suppliers involved in the development projects could be registered elsewhere in China. If the Qianhai-based firm is also related to or made arrangements with its supplier(s), then money can be channelled through easily. Second, once the loans have been actualised, they need not be immediately invested in the construction projects. This raises the possibility that the funds could be directed to other uses. So long as the borrower fulfils interest payments, no alarm bells would ring, not in the short term at least.

Another major issue lies in whether the new policies could jumpstart interest rate liberalisation onshore. Shortly after the first-wave cross-border loans from Hong Kong, the *People's Daily* (29 January 2013) termed the reforms in Qianhai 'interest rate liberalisation accelerated'. Interest rate controls are integral to the CPC's ability to launch independent monetary policy. To be sure, 'independence' in the Chinese context is a paradox: the PBoC is embedded within the party-state apparatus, which means its strategies are inevitably extensions *of* rather than truly independent *from* political considerations. In addition, the four biggest commercial banks in China are all state-owned and are responsible for loans to state-owned enterprises, further blurring the notion of 'independence' between the CPC and the financial system.

A new contradiction emerged in July 2013 when China-based banks were allowed to lend at rates that they wish. While this technically meant interest rate 'liberalisation' has occurred in the Chinese financial system, its impact is expected to be minimal since the majority of loans continue to be priced at or above the PBoC-set benchmark rate (*New York Times*, 21 July 2013). In other words, even with nominal 'liberalisation', so long as the commercial banks (most of which are state-owned or state-linked) remain entwined in a state-governed system involving the PBoC, the CPC could still directly influence interest rate-setting and credit creation. Because of this multi-layered entwinement between financial institutions and the party-state apparatus, it is hard for one aspect of reform to set in motion a systemic overhaul. On the contrary, this development reinforces the possibility that policy changes were made to enhance CPC control of the existing regulatory system.

It is in relation to private sector economic activities that the PBoC retains true 'independence'. Through interest-rate targeting and another major macroeconomic policy, capital controls, the PBoC is able to strongly influence the supply and allocation of private capital within China. And it is arguably this 'independence' that is subject to regular critiques by the international community. The former US Treasury Secretary, Henry Paulson, framed 'interest rate liberalisation' in China as a 'necessity' shortly after the designation of Qianhai New Area: 'What we need – and the people I talk to in Beijing understand this – is interest rate liberalisation. If it's not done early, it sends a very negative signal' (*The Wall Street Journal*, 6 June 2013).

Understanding is different from acting, however. The Chinese government has shown clearly through its unrelenting grip on both the existing macroeconomic regulatory mechanisms and the barriers to entry into the banking industry that the domestic financial system remains off-limits. Ironically, it could be because of this backdrop of persistent state intervention in the financial industry that the Qianhai reforms sent positive signals that China was about to institute unconditional interest rate liberalisation. After all, it was the first time that firms in the Chinese mainland could benefit from competitive lending rates truly independent from PBoC-influence. These signals in turn raise a broader question regarding structural reforms: if interest rates in the mainland economy would increasingly be determined by the market (hence the 'liberalisation' mentioned above), does this mean that there is an imminent attempt to either remove the fixed exchange rate regime or effect free capital flows?

As the policies in Hengqin already suggest, the financial reforms in the GPRD do not undermine the existing national institutions of fixed exchange rate and capital controls (ref. preceding section). On the contrary, these reforms could reinforce the continuity of these institutions. Nationwide interest-rate liberalisation could actually generate two contradictions for Qianhai as an RMB backflow channel. First, a fully 'liberal' interest rate regime would mean no a priori necessity for firms on mainland China to borrow from Hong Kong-based banks – firms could obtain loans from any bank in the world that offers the best interest rate. Interest-rate liberalisation would thereby undermine Hong Kong's role as an offshore RMB centre. Second, interest rate liberalisation involving both onshore and offshore lenders could undermine one of the two other forms of financial regulatory tools currently employed by the CPC, namely the dual institution of capital account control and the fixed exchange rate regime. Not only would the CPC lose the capacity to set interest rates, capital could flow to offshore RMB centres that offer higher interest rates, in turn undermining the existing control on portfolio in- and outflows and, by extension, the RMB exchange value.[5]

Further observations suggest deeper structural changes will not take place in the short run, at least not via the reforms in Qianhai. Speaking to *Reuters* in 2013, an unnamed Hong Kong-based banker made it clear that current cross-border loans to Qianhai-based may reach its geographical limits if no further modifications are made:

> Infrastructure construction in Qianhai is the first stage for loans and after that, you'll have to allow broader use of these loans to facilitate operations and trade. After all, Qianhai is a very small area in Shenzhen, just like Central in Hong Kong. A company can set up an office there, but its factories in other mainland cities also need financing. (*Reuters*, 15 August 2013)

To Jin Xinyi, a member of Shenzhen's Chinese People's Political Consultative Conference, Qianhai plays a distinct economic–geographical function:

> The Qianhai project is an experimental economic zone with the clear goal of making the yuan a global currency, instead of a pilot project for administrative and legal reforms...Institutional reform in Qianhai will only focus on those fields that promote the city's modern service industry and industry upgrades and enhance cooperation between Shenzhen and Hong Kong. Hopes are slim that concrete administrative and legal reforms will follow in Qianhai. (*South China Morning Post*, 11 May 2013)

In addition, while the Qianhai-Hong Kong flows are supposed to be two-way, the other direction of flow – namely the ability of Qianhai-based firms to invest directly in the Hong Kong financial markets without regulation – remains subject to the central government's control (ref. Table 5.2). This control in itself is an encumbrance to deeper financial integration in the GPRD. In late November 2013, the Hong Kong Financial Services Development Council (FSDC) proposed to the Hong Kong government to lobby Beijing to approve a program dubbed 'Qualified Domestic Institutional Investor 3' (QDII3). Building on the preexisting QDII program that allows selected mainland Chinese firms to make portfolio investments abroad, the FSDC (2013: 6) would like Qianhai-based investors to be able to make up to 50 billion RMB-denominated investments (~US$8.25 billion) and up to 5 billion dollar-denominated investments per year.

As the FSDC (2013: 5) explains, there was a pressing need to enhance the flow of RMB into Hong Kong: 'Most financial institutions which engage in the renminbi business in Hong Kong consider the lack and instability of renminbi liquidity the biggest bottleneck of the offshore renminbi market'. Going by the observation of an experienced Hong Kong-based venture capitalist, the cautious approach of the Chinese government may mean it would be quite a while yet (if at all) for this 'bottleneck' to be removed:

> This scheme is great, it will attract many financial firms to apply to move to Qianhai in order to enjoy access to the Hong Kong financial market. Going by the way things are done, there is a good chance that, once approved, the quota will go up. But Beijing has still not approved the QDII2 scheme [an earlier proposal to allow off-shore portfolio investments from Guangzhou and Shenzhen]. They said they will approve it in 2012, and we still have not heard anything. QDII3 will not become reality so soon. (Email correspondence with venture capitalist, 15 December 2013)

This observation proves prescient because QDII3 remains a proposal in the spring of 2018 (the time of writing). The 'slim hopes' of concrete administrative and legal reforms and the limitations of the first-wave cross-border loan reforms jointly exemplify the tensions intrinsic to state rescaling, policy experimentation and path-dependency as discussed in Chapter 3, namely that between a subnational developmental agenda (the Wang Yang administration's attempt to attract higher order industries, as outlined in Chapter 4), national-level structural coherence (a desire to retain the quasi-insulated financial system) and transnational circulatory capital (in this context flowing primarily through Hong Kong). For this reason, the establishment of these onshore enclaves of offshore RMB investments could never guarantee success a priori: it is contingent on whether offshore investors are confident in profiting from the backflow channels identified by the CPC. Furthermore, as the same Hong Kong-based venture capitalist mentioned in an earlier interview in Hong Kong, Qianhai could be a risky spatial strategy to enhance the functions of state-owned banks based offshore:

> Actually this new development is very interesting, several banks in Hong Kong with large deposits are in fact subsidiaries of mainland Chinese banks, under the current regulations these subsidiaries cannot direct these funds into private equity in China. Something must be done to allow them to use these deposits, and Qianhai is the first step. But how would Qianhai stand out compared to other parts of China? If it's only the possibility for some offshore RMB to come back legally into China, then its advantage is temporary. Offshore RMB can go anywhere and everywhere so long as the state permits it. Or if the RMB becomes fully convertible. In terms of investments, right now it seems to be centred on real estate, I'm not sure if Qianhai will end up like another 'hollowed out' real estate project in China. (Interview, venture capitalist, Hong Kong, January 2013.)

This observation corresponds to the earlier comment of the limits to cross-border lending by the unnamed Hong Kong-based banker: if no new avenues for capital flows are created, pre-determined new state spaces like Qianhai would be quickly saturated by fixed capital investments. At the local level, this saturation could easily be offset by the identification of new 'spatial fixes' – RMB backflow channels could indeed be scaled 'anywhere and everywhere' across Chinese state spatiality. More importantly, these new spatial fixes would guarantee a short-term boost to GDP. At the structural-systemic level, however, if capital is not put into more 'productive' use beyond the financial and real estate industries that could generate longer-term revenue streams, it is unclear if offshore investors would find it attractive to see their funds channelled to the already overheated housing markets and infrastructural overcapacities (e.g. underused freeways and airports).

Fundamentally, the intention to plan for rather than allow markets the freedom to determine the geographies of RMB backflow channels underscores the necessity – if not centrality – of state involvement in the Chinese financial system. This may appear paradoxical vis-à-vis the demands for global capital integration and its supposed corollary, policy convergence, but it is arguably necessary to sustain the CPC's commitment to Deng's Four Cardinal Principles (ref. Chapter 1). Ji Zhihong, Director of the Research Bureau of the People's Bank of China (PBoC), puts this necessity in historical perspective:

> Historically there was a tradition of privately-operated banks, for instance prior to liberation [i.e. 1949] there were large numbers of money and credit houses. But under the implementation of a highly-centralised planned economic system prior to reforms in 1978, banks were just like fiscal monetary pockets. Naturally, under such a system without a privately-operated economy, there was no need for privately-operated banks. Even though the [post-1978] reforms have gone on for so many years and the privately-operated economy has developed and strengthened, at a certain level, banks are still viewed as necessary tools to determine the economic pulse and facilitate macro-level adjustments...In reality, to look at it from many years of social debates, the issue regarding privately-operated banks is at a certain level a problem with the state's attitude towards fair competition. (Ji, personal essay published in *Caijing*, 16 June 2013; author's translation)

It is in relation to the central government's persistent desire for control over the Chinese economy and yet enjoy the privilege of a global currency that the national designation of Qianhai was conceptualised. For the RMB to be truly 'internationalised', it must have the ability to circulate strongly outside of the borders of its 'home economy'. According to received textbook economic theory, the precondition of currency internationalisation is the freedom of the international currency (e.g. the US dollar, the Japanese yen and the euro) to come and go freely from its respective 'home economy'. The introduction of experimental policies for cross-border loans in Qianhai at once reinforces and revises this theory: it is reinforced by the CPC's acceptance of the requirement for the RMB to move freely across geopolitical borders, but revised by the restriction of this freedom to spaces outside of mainland China. Through instituting offshore RMB centres around the world, RMB can flow freely in and out of these centres unencumbered at a quota set by the CPC. Unlike other internationalised currencies, however, these RMB flows would not be able to flow freely into China, its 'home economy'. The central reason is because the CPC wants to retain the power to shape the geographies of these backflows, and following active lobbying by Guangdong's provincial-level policymakers, Qianhai New Area became the economic–geographical expression of this power.

Conclusion: Change as the Precondition of Continuity?

The designation of Hengqin and Qianhai as 'nationally strategic new areas' is at once a local initiative and a national project that is simultaneously related to the function of other analogous projects within which the central government holds significant leverage. It is an expression of what Peck (1998) terms 'central localism', a mode of governance in which the central government of a nation-state identifies and capitalises on local socioeconomic characteristics to attain its goals. However, as Chapters 4 and 5 illustrate, these new areas did not emerge from a geographical–historical vacuum: they came into being in part through the Guangdong government's risk-laden attempt to effect the 'double relocation' restructuring policy following the global financial crisis, and in part through the Chinese central government's concomitant desire to reduce its use of the US dollar after the crisis exposed the instability of the US financial system. In one sense, then, in as much as the state rescaling process and the consequent policy innovations represent unprecedented changes, they are nonetheless *still* embedded within a broader system that has deeper historical roots.

A major path-dependent aspect of this system – a legacy of the 'cellular' economic structure discussed in Chapters 1 and 2 – is the lack of integration between local governments within the GPRD. As Chun Yang (2005: 2147) has observed, 'Decision-making competencies in the HK–PRD cross-boundary region have tended to disperse across multiple levels of governments...In response to the lack of an effective regional authority and the unique political framework of "one country, two systems," [the] central government has played a somewhat backstage coordinator role in the transition of the integration mechanism and cross-boundary governance'. A former participant in discussions on the planning of Hengqin, Wang Xianqing (2010: n.p.; author's translation), made a similar observation of the history of inter-governmental collaboration in the PRD: 'It can be said that, all over the country, there is no region like the PRD when it comes to lack of collaboration [between local governments], unwillingness to collaborate, or being plainly difficult during collaborations. These local governments make meticulous calculations based on very short-term interests, they seek instant benefits (*jigong jinli* 急功近利), and they are simply unwilling to cooperate sincerely and magnanimously on the basis of a broader strategy'.

In view of this 'lack of an effective regional authority' and the local governmental culture of being 'plainly difficult' when it comes to collaborative strategising, Wang Yang arguably found it both necessary and advantageous to involve governments positioned at other scales – namely the Chinese central government and the Hong Kong and Macau SAR governments – in the spatial repositioning of Guangdong. Embedded within this repositioning strategy was the attempt to 'scale up' three intra-urban territories – Hengqin, Qianhai and Nansha – in order to retain the economic importance of Guangdong province (ref. Figure 4.1). The underlying logic was to facilitate fresh engagements with transnational circulatory capital just as industries deemed redundant were either relocated to the outlying parts of Guangdong or were allowed to expire automatically without further state support.

Through this two-pronged spatial strategy, the Wang administration simultaneously set in motion and responded to the reconfiguration of the provincial economic–geographical structure. This proactive role of the Guangdong government in driving the rescaling process corresponds with Brenner's (2004: 93) observation that '[s]tate institutions do not contain a pre-given structural orientation towards any specific scale, place or location; however, determinate forms of state spatial and scalar selectivity emerge insofar as social forces successfully mobilise state spatial strategies that privilege particular spaces over others'. By extension, it further supports the conceptual argument raised in Chapters 1 and 3 that experimental reforms in post-Mao China are not underpinned by absolute regulatory transfers to local governments. Rather, the reforms exemplify a process of 'decentralisation as centralisation', or the enhancement of central state power through the introduction of 'nationally strategic' policies in targeted locations (predominantly at the intra-urban scale), with centrally-owned state-owned enterprises (SOEs) playing active roles within these newly designated territories.

As this chapter has discussed, however, it is still early to know if the production of these experimental economic spaces will lead to long-term national structural coherence. Indeed, it is a tentative, risk-laden process generated by the CPC. As one Macau-based legal consultant puts it, 'it appears that they [CPC-linked actors] are just playing amongst themselves' (Interview, January, 2013). Even before offshore funds entered Qianhai and Hengqin New Areas, the colossal capital inflows into these areas were supplied primarily by debt-financing from the central to the provincial levels. While the inflows had the practical effect of easing the political opposition to and the economic effects of the 'double relocation' program, it arguably worsened the national debt-to-GDP ratio and further underscores the sustained reliance on debt-financing and investment-driven growth in the Chinese macro-economy. Whether the invested funds could be recouped is thus an open question. It also remains to be seen if these offshore funds could penetrate the previously-mentioned 'concrete economy', namely sectors that contain opportunities for expanded capital accumulation beyond the financial and real estate sectors.

To examine how the state rescaling process contributes at once to regional economic integration and national-level economic restructuring, this chapter documented the rationale of the experimental financial reforms in Hengqin and Qianhai. It can be argued that these 'new areas' are functional extensions of new policies introduced in Hong Kong and Macau. Underpinning this functional extension is territorial re-bordering, institutionalised as 'one- and two-line borders' in Hengqin and through the regulation of cross-border credit flows to and from Qianhai. These territorial re-designations have added new dimensions to the economies of Hong Kong and Macau, and in turn illustrated specific issues confronting the central government as it works to reconfigure the national financial structure. Through this analysis, new questions are asked of the national-level institutional foundations (cf. research agenda, Chapter 3): are the policy experiments to facilitate more market-like governance and a concomitant 'retreat' of the state, or, perhaps counter-intuitively, could they be intended to reinforce the CPC's involvement in the economy and in turn enable it to remain the predominant driver of capital accumulation across Chinese state space?

Empirical evidence suggests this round of state rescaling in the GPRD allowed the CPC to enhance its control of the national financial system. That this process involved a series of policies aimed at deepening cross-border economic integration with Hong Kong and Macau accentuates the importance of economic liberalisation in the Chinese central government's calculations. The impetus to deepen experimentation with market-driven development supports Brenner's (2004: 111) contention that 'state institutions must be conceived as multi-scalar sociospatial configurations that evolve historically, often in ways that have significant ramifications for the geographical configuration of capitalism as a whole'. At the same time, however, the 'new areas' also represent the *politico-geographic limitations* of economic liberalisation – just like it was in 1978, the Chinese central government remains in the mode of 'feeling for stones' as it crosses the 'river' of liberalisation (ref. Chapter 1). Rather than signal an inevitable movement towards capital account convertibility and interest rate liberalisation, the reforms are in fact beginning to show how the Chinese government wants to have its cake (i.e. to reduce its exposure to the risks of using the US dollar) and eat it (i.e. to expand the global usage of the RMB through an 'internationalisation' drive). Perhaps the most important finding of this chapter is the retention – if not also reinforcement – of Chinese state power in directing offshore capital investments to onshore locations (Qianhai in this regard, although Hengqin could play the role if the 'second line' border begins to institute currency controls; ref. Figure 5.3).

Through illustrating how central government power is facilitated through its experimental approach at economic restructuring, this chapter provides analysis for examining the historical connection between national-scale structural coherence and institutional reforms. With reference to Figure 3.1 (Chapter 3), the reforms instituted in Hengqin and Qianhai New Areas illustrates how the strong regulation of transnational circulatory capital remains the sine qua non of national structural coherence. It could indeed be argued that, despite rising debt-to-GDP ratios and constant threats of massive local governmental defaults on loans, the Chinese financial system has remained stable *because* foreign capital had been unable to generate systemic damage last seen during the 1997 Asian financial crisis. The latest series of policy experimentation in and through the GPRD thereby underscores the book's overarching conceptual argument that the tensions between state rescaling and path-dependency have generated qualitatively distinct outcomes that may be shifting, but certainly not undermining, existing regulatory foundations.

Endnotes

1   The official term of this cross-border strategy in Mandarin is <关于全面推进金融强省建设若干问题的决定>.
2   Within the Chinese scalar hierarchy, Hong Kong and Macau SARs are ranked as quasi-foreign entities. This means the SAR governments' negotiations with Chinese provincial authorities must formally involve the Chinese central government, technically its scalar equivalent (although officially the central government could veto decisions made by the SAR governments).
3   Zhou was the primary proponent and driver of the RMB internationalisation drive since 2009. The Qianhai spatial strategy thus dovetailed with and became an exemplification of his plans: it became the first site to test the creation of 'backflow mechanisms' that would in turn enable the Chinese government to retain the fixed exchange rate regime (more on this in the next section).
4   RMB-denominated trade is a two-way process and includes the use of documentary credits such as letters of credits from the designated economies. On the mainland side, not all firms can participate in this trade; firms must first be approved as Mainland Designated Enterprises (MDEs) and must be based in specific cities (four of the initial five were in Guangdong, namely Guangzhou, Shenzhen, Dongguan and Zhuhai; Shanghai was the other city). To open a trade account, foreign companies must demonstrate genuine transactions with the MDEs.
5   This was the process that undermined the entire Thai economy within a very short period in 1997 and that triggered the Asian financial crisis.

Chapter Six
State Rescaling in and Through Chongqing I
The State as Economic Driver

Introduction

Beginning from 2007, the CPC implemented a series of experimental socioeconomic reforms in Chongqing, a city-region[1] in southwestern China (ref. Figure 6.1). While reforms have been commonplace in the post-Mao era, the new policies in Chongqing earned a reputation as the 'Chongqing model' or 'Chongqing experience' of development because it (i) reinforced strong state involvement in economic development, leading to the designation of Liangjiang New Area as the first 'nationally strategic' growth pole in western China (the preceding two were Binhai New Area in Tianjin and Pudong New Area in Shanghai); (ii) expanded access to social welfare for rural migrants in the core urban areas; and (iii) mobilised the masses through a series of propagandistic campaigns. The latter process, driven by Chongqing Party Secretary Bo Xilai between 2007 and 2012, was symbolised through a recurring series of 'red song' singing campaigns (*changhong* 唱红) reminiscent of the Mao-era. And just like the Mao-era campaigns to publicly humiliate 'class enemies' and 'counter-revolutionaries', Bo gave massive publicity to his programs to eradicate organised crime (*dahei* 打黑). Newspapers regularly reported the arrests and swift prosecution of triad members, and a museum was established at the most iconic location of the municipality – Chaotianmen – to showcase the faces of key arrested gangsters.

On Shifting Foundations: State Rescaling, Policy Experimentation and Economic Restructuring in Post-1949 China, First Edition. Kean Fan Lim.

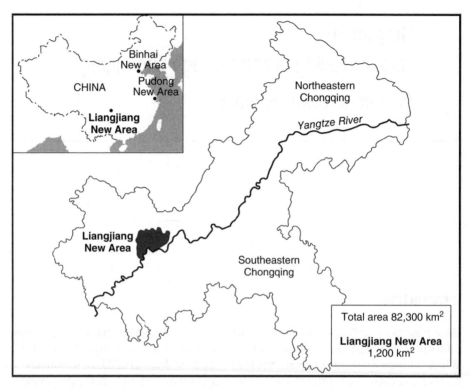

Figure 6.1   Location of Liangjiang new area within Chongqing, viewed in relation to the locations of Binhai and Pudong new areas (inset, top left). Source: Author, with cartographic assistance by Elaine Watts.

It was arguably for this reason that the reforms in Chongqing were construed as a concoction of terror-like campaigns and a cult of personality characteristic of the Cultural Revolution (1966–1968). In March 2012, just over a month after Bo was implicated in a corruption-cum-political scandal, the-then Chinese Premier Wen Jiabao made a veiled critique – broadcasted live on national TV – of reforms in Chongqing (CCTV, 14 March 2012; author's translation). Shortly thereafter, influential actors began to make concrete associations between the Chongqing reforms and the Cultural Revolution. Hu Deping, the vice minister of the Propaganda Department of the CPC, argued that both during the Cultural Revolution and in Chongqing under Bo, 'countless wronged cases occurred in both instances under the prettiest discourses, with people subjected to disaster' (*China Reform*, 4 December 2012). Xu Youyu (2013), a prominent Chinese academic, similarly argues that, even though the Chongqing reforms were taking place in a globalising context, they were highly similar to the Cultural Revolution because (i) the rule of law was suppressed; (ii) ideological motivations were

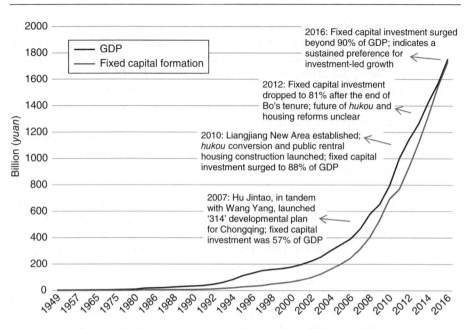

Figure 6.2 Fixed capital investments and GDP in Chongqing, 1949–2016. Source: Chongqing Municipal Bureau of Statistics (2017). Author's graphical illustration.

explicit; and (iii) one segment of the population offered significant support. While the socio-ideological mobilisation programs in Chongqing stopped after Bo was removed from power in March 2012, a 2017 inspection by the Central Commission for Discipline Inspection (CCDI) found that the 'pernicious ideological legacy' left by Bo and his right-hand man, Wang Lijun, 'had not been wiped out completely' (*South China Morning Post*, 14 February 2017a).

Interestingly, however, fixed capital formation as a share of GDP grew beyond 90% after Bo's tenure (see Figure 6.2). Annual GDP growth remained above 10% at last count in 2016, while the flagship economic project of the Bo era, Liangjiang New Area, has evolved into a thriving industrial hub and the base of China's third government-to-government (G-to-G) developmental project with Singapore. These developments collectively exemplify the consolidation of a path-changing, investment-driven approach that preceded Bo Xilai's tenure as Chongqing Party Secretary. Indeed, the socioeconomic reforms undertaken by Bo were set within parameters of the '314' developmental plan established by his predecessor, Wang Yang. As Chapters 4 and 5 have shown, Wang was the Guangdong Party Secretary who went on to launch the 'double relocation' economic–geographical restructuring in the province. What Bo and his successors – Zhang Dejiang and Sun Zhengcai – did was therefore to build directly on the path established by Wang rather than resuscitate a blueprint instituted during the Mao era. Specifically, Bo embellished Wang's agenda with a program of export-oriented industrialisation that culminated in the inclusion of Liangjiang New Area within regional and global production networks of transnational enterprises. The process was politicised, however, because it was entwined with Bo's explicit emphasis on socialist development (which led to the associations with the Maoist developments mentioned earlier). Underpinning the designation of Liangjiang New Area are therefore *politics* that reflect the tensions between state rescaling, policy experimentation, and path-dependency as outlined in Chapter 3. The joint aim of Chapters 6 and 7 is to foreground, conceptualise and evaluate these tensions.

Chapter 6 presents a history of the emergence of strong state intervention in Chongqing, and explains how the designation of Liangjiang New Area extended this interventionist legacy. Chapter 7 documents and evaluates the tensions associated with socioeconomic reforms in the municipality. Following Jun Zhang and Jamie Peck (2016), the chapters point out that the 'Chongqing model' cannot be reduced to the *changhong* and *dahei* programs. In addition, they do not represent the work of a single political actor. Indeed, the development of Chongqing has its own history, which means it is never just a microcosm of what is occurring at the national scale. Situated within the analytical framework presented in Chapter 3, Chapters 6 and 7 will explain (i) how subnational and transnational actors interact to produce the conditions that made possible the national designation of Liangjiang New Area' and (ii) how this designation is entwined with an experimental attempt to reform the national-level 'urban rural dual structure'. *Collectively, these chapters argue that the reforms in Chongqing appear to exemplify (Maoist) path dependency because of the rolling effects of policy experimentation in the post-Mao period.*

This chapter is arranged in three parts. The next section will examine the relationship between proactive state intervention in Chongqing (a primary target of contemporary critiques) and changes in its regulatory relations with the central government. It argues that the strong governmental involvement reflected variegated developmental paths generated by spatially-selective policy experimentation during the first phase of market-oriented reforms (i.e. the 1980s). The third section will then demonstrate why the introduction of experimental regulatory capacities to Liangjiang New Area was the outcome of active political lobbying in the post-crisis period amidst intense competition from other cities. This section further reinforces the primary argument presented in this book: state rescaling is at once an expression of subnationally-driven developmental agendas in the name of the 'national interest' and an effect of developmental pathways instituted by previous regulatory regimes. The concluding section discusses the conceptual implications of the empirical findings, giving emphasis on the importance of evaluating the 'Chongqing experience' across a broader geographical-historical context.

THE STATE AS ECONOMIC DRIVER 149

The Evolving Positionality of Chongqing: A Geographical–Historical Overview

Chongqing is one of the four province-level cities under the direct control (*zhixiashi* 直辖市) of the Chinese central government (as introduced in Chapter 2). It assumed an important economic–geographical position within China since modern industrialisation began towards the end of the nineteenth century. Developed into imperial China's first inland open port in 1891, connections with Britain, France, Germany, and Japan were established by the turn of the twentieth century. Chongqing went on to become the primary trading centre along the upper Yangtze River (for a detailed history of this period and earlier, see Zhang et al. [2015] and Chongqing Research Academy of Education Sciences [2015]). This pre-existing economic position, coupled with its strategic location behind several mountain chains far from potential naval landings by foreign armed forces, were arguably the reasons why the Kuomintang (KMT) selected the city region as the temporary capital of the Republic of China[2] during World War II (1937–1945) and the Chinese civil war (1945–1949).

A colossal amount of capital goods and labour power was relocated from other parts of China to Chongqing during this wartime period. This produced a vast military–industrial complex in the municipality and further enhanced its position as a major economic centre. While the Communist Part of China (CPC) removed Chongqing's province-level status and merged the city with Sichuan province after it became the official government of China in 1949, Mao would, interestingly, repeat what the KMT did in Chongqing. In 1964, Mao declared China was facing massive geopolitical threats and decided to launch what was introduced in Chapter 1 as the 'Third Front Construction'. Under this program, China's coastal belt was designated a 'first front' of defence, and the southwestern interior as the 'third front'. All the areas in between are categorised as the 'second front'. Many industrial and military complexes were consequently relocated to or constructed in the city-region, adding a new layer of industries and labour power to an industrial structure well-established by the KMT.

As mentioned previously, the Third Front construction was not the first instance through which Mao-era regulatory logics re-expressed regulatory logics of the KMT era. This development is of theoretical significance for conceptualising path-dependency in post-1949 China: logics inherited from earlier regimes were not totally jettisoned but repurposed to suit new regulatory contexts. The Mao administration combined selected techniques used by the KMT with new policies, in turn generating new developmental pathways (e.g. collectivised production, rural industrialisation, the re-concentration of heavy industries in inland provinces, etc.). This approach would similarly characterise post-Mao development, where Deng and his successors repurposed selected Mao-era regulatory logics (e.g. land nationalisation and the hukou system of demographic control) and combined them with new experimental policies in targeted territories along the eastern seaboard. It is this dynamic interaction between the past and the present that defined Chongqing's emergence into a major frontier of reforms in contemporary China.

Prior to this emergence was a steep decline in Chongqing's economic–geographical significance during the initial post-Mao era (i.e. between the 1980s and the early 1990s). As Chapters 1 and 3 have explained, the CPC tried to retain the structured coherence of the national economy through the 'ladder-step' logic. This logic was first expressed in 1980 through the formation of four Special Economic Zones (SEZs) in China's southeastern region (as introduced in Chapter 1), before expanding into a full-fledged coastal-oriented 'industrialisation with urbanisation' phenomenon in the early 1990s. For more than two decades, then, Chongqing's economic growth – like many cities and provinces in China's western interior – lagged provinces along the coast. National income inequality worsened correspondingly.

It was only after the Three Gorges Dam national energy project was launched in the mid-1990s that Chongqing re-appeared on the map as a strategically significant city. Because of the sheer size and scale of the dam construction, many settlements in the upstream valleys of the Yangtze River were inundated. An estimated two-thirds of the eight million affected residents were originally located in Sichuan province, the majority of them living in poverty. This engendered a colossal logistical and governance issue: the Sichuan provincial government, which already had a huge and relatively poor population under its charge, was confronted with the challenge of resettling and then re-employing the affected residents without further straining its administrative and fiscal resources.

Proposals to create a new province in the Three Gorges region began to appear at the time.[3] It soon became apparent, however, that the central government was keen to circumvent the costly administrative duplication and politics associated with establishing a new province. According to a 1986 diary account by the former Premier, Li Peng, central policymakers were primarily concerned with prioritising the dam-building project over the redrawing of provincial boundaries (Li 2003). Since Chengdu, the provincial capital of Sichuan, was too far away from the flooded areas, Chongqing was identified as a more feasible location for the population resettlement issue. To ease the financial impact on the Sichuan government, the central government agreed to administer the districts of Fuling, Wanxian and Qianjiang, then amongst the poorest areas in Sichuan, as resettlement areas. These districts were then merged with Chongqing's core urban zone to form the sprawling Chongqing Municipality. It is intriguing to see in the Chongqing experience history coming full circle: in an almost identical repetition of what the KMT regime did in 1937, the CPC 'promoted' the city-region into a centrally-governed municipality for the second time in 1997.

Arguably because of its relative (in)significance to the initial market-oriented reforms, Chongqing policymakers were impelled to enhance the capacities of its

SOEs to drive economic growth. As one Chongqing-based academic from Sichuan explains, this phenomenon expresses a two-way connection between economic–geographical inequality and SOE reforms:

> The reforms of SOEs not only affected regional inequality, you can say it was forced by this inequality. The rich coastal regions had clear advantages over the poor inland regions during the reforms of SOEs. Because the private sector was growing in the coast, it was relatively easier to reemploy laid-off workers from SOEs. This is a major political problem for all municipal governments, these workers won't accept going to the countryside! Privatisation or restructuring SOEs was also easier along the coast because capital and labour markets were more developed. The biggest problem, for me, is the concentration of many large SOEs in the inland regions. Employing so many people and yet lagging behind the coast economically, it was hard to restructure or remove these SOEs directly. (Interview, Chongqing, March 2012; author's translation)

It was arguably for this reason that the state-driven industrialisation approach was not fully dismantled in the post-Mao era. Since the late 1980s, some SOEs involved in weaponry manufacturing were relocated; some were fully abandoned; while others like Ansteel, the Jialing Group and the Changan were restructured and are now dominant SOEs based in the municipality. In addition, the Chongqing government actively encouraged SOEs to list in foreign capital markets. Building on the SOE restructuring is the 'Eight Major Investments' (*badatou* 八大投) program in infrastructural construction by the Chongqing government. This program involves the establishment of eight state-financed corporations, each investing in functions within sectors identified as fundamental for industrialisation (see Table 6.1).

Because of the proactive restructuring of state involvement in the economy, the Chongqing government increased the value of state-owned assets (excluding those of central government-controlled SOEs) from 150 billion *yuan* in 2000 to 1.8 trillion *yuan* in 2013, a 12-fold increase (Huang 2006: 20; *Caijing*, 20 January 2014). In an intriguing twist of fate, what were originally geographical–historical liabilities in the post-Mao era intriguingly transposed into economic assets. As the next section and Chapter 7 will demonstrate, it was this strong state capacity, duly repurposed, that jumpstarted what became widely referred as the 'Chongqing experience' of reforms.

Chongqing as a New Platform for Policy Experimentation

The Path of State Rescaling in and Through Chongqing

As the framework in Chapter 3 shows, state rescaling occurs dynamically in relation to developmental goals at both the subnational and national levels. In Chongqing, a new round of spatially-targeted policy experimentation was set in motion after the

Table 6.1 The firms and functions of 'eight major investments' in Chongqing.

Firm	Function
Chongqing Expressway Development Co. (CEDC)	Constructs, operates, and manages expressways.
Chongqing Transportation and Tour Investment Co. (CTTIC)	Constructs, operates, and manages highways; develops and manages tourism attractions.
Chongqing Urban Construction Investment Corporation (CUCIC)	Develops urban infrastructure such as bridges, tunnels, and roads in the main urbanised districts.
Chongqing Energy (Construction) Investment Corporation (CEIC)	Invests, operates, and manages energy-related (i.e. electricity, gas, and coal) power projects
Chongqing Real Estate Group (CREG)	Restores and develops lands.
Chongqing Development Investment Corporation (CDIC)	Builds and operates rail transportation and other infrastructure projects.
Chongqing Water Works Controlling Group (CWWCG)	Provides water supply and drainage integrated service to the main urban area.
Chongqing Water Resources Investment Co. (CWRIC)	Invests in and constructs water conservancy projects, small hydropower plants, and water supply and drainage projects.

Source: Author's compilation.

global financial crisis struck in 2008 offered an excellent context to step out of established developmental pathways. This point needs elaboration: the crisis may have appeared as an external shock, but the conditions and intentions for experimental reforms were established prior to the crisis. Interestingly, as discussed in Chapter 4, this was similar to the Guangdong government's efforts to cope with the financial crisis. In turn, it lends further support to the conceptual argument in Chapter 3 that state rescaling is not a singular expression of decentralised governance or recentralised state power – it reflects the *interactions* of institutions, policies, and stakeholders across different scales.

While the Chongqing government was confronting the challenge of restructuring SOEs through the 1990s, it was becoming apparent that the enlargement of Chongqing municipality had exacerbated urban–rural income disparities. While Chongqing is nominally a municipality, more than two-thirds of its population held rural *hukous* after the municipality merged with the Three Gorges regions. Shen Xiaozhong, the Director of Chongqing's Municipal Development and Reform Commission, puts the situation in perspective:

In 2006 the urban–rural income ratio in Chongqing was 4 : 1, it was larger than the disparities at the national and western regional level; the GDP ratio between in core city area, western Chongqing and the Three Gorges watershed area was 3 : 1.6 : 1, it was greater than the ratio between the eastern central and western parts of the country; the difference between the per capita GDP was 10.5 times, it was larger than the difference between Shanghai and Guizhou. In addition, the resettlement of millions of people in the Three Gorges area was a world-level challenge, be it in human resources or tasks it was more complicated than any other construction project nationwide. (Shen, interview with *21st Century Herald*, 6 April 2007; author's translation)

As Shen adds, this disparity illustrates the contradictions of the Mao-era 'dual structure', and was thus the target of reforms:

> The holistic reforms on urban–rural integration and the consequent experience have a demonstration effect at the national level. The pronounced urban–rural dual structure is a microcosm of a basic national condition. As a special intersection of a big city, a big countryside and a big watershed, in addition to its location in the interior, the exploration of institutional reforms in Chongqing serves as a model for provinces in the Chinese interior. (Shen, interview with *21st Century Herald*, 6 April 2007; author's translation)

Yet there had to be more specific reasons why Chongqing was suitable as a site of reforms vis-à-vis other locations that were similarly subject to the constraining effects of the 'dual structure' (ref. comment by Qiu Feng in the introductory section). The interview with the Chongqing-based planner indicates the intra-municipal disparity was ironically exacerbated by the Three Gorges resettlement program (ref. discussion of how this structure was undergirded by the *hukou* institution in Chapter 2):

> As a heavily industrialised city, a significant portion of people in Chongqing enjoyed employment stability and good social benefits, so when a large number of poor rural residents were categorised as Chongqing residents, the urban–rural income disparities in Chongqing naturally became larger. And the government suddenly had a new responsibility to reduce this disparity. So while the city [in its much smaller, pre-1997 form] struggled to catch up with the coast, it also had to overcome this disparity. (Interview, Chongqing, March 2012; author's translation)

Viewed in relation to Shen's comment, this insight indicates the legacy of the 'dual structure' was not simply a 'microcosm' of 'a basic national condition': it was 'pronounced' in Chongqing because of the *interaction* between the 1958 *hukou* institution, the 'Third Front' industralisation drive and the Three Gorges Dam construction project. The region-specific interaction of these inherited policies was arguably the primary reason why the Chinese central government, then led by Hu Jintao, selected Chongqing as the site to experiment with removing this 'dual structure'.

Facilitating the 'demonstration effect', in Shen's parlance, was the '314' integrated developmental strategy. '3' in '314' refers to three major positions, namely (i) to accelerate Chongqing's position as a 'growth pole' (*zengzhangji* 增长极) in western China; (ii) to be the economic hub of the Yangtze upstream; and

(iii) to develop through urban–rural integration (*chengxiangtongchou* 城乡统筹). '1' in '314' refers to the broad overall objective of attaining the first 'affluent society' in western China, while '4' in '314' refers to '4 major tasks' (*sidarenwu* 四大任务), namely to (i) expand 'industrialisation to stimulate agrarian development' (*yigongcunong* 以工促农), 'lead the countryside through the city' (*yichengdaixiang* 以城带乡) and develop a new 'socialist countryside'; (ii) realistically transform the means of economic growth and accelerate the reformation of old industries; (iii) address the livelihood problems of the people so as to develop a Harmonious Society; and (iv) holistically enhance city-building and raise the standards of urban management.

Launched in March 2007 by Wang Yang (who went on to lobby for the 'nationally strategic' designation of Hengqin, Qianhai, and Nansha New Areas in the GFTZ) and the-then Chinese President, Hu Jintao, this strategy was the first instance Chongqing was identified as a 'growth pole' of the western interior (see detailed description in Table 6.1). It was also the first instance the Chinese government explicitly sought to tackle the problem of treating rural migrants in the cities as citizens with equal social rights as those holding urban hukous. After Bo Xilai was appointed the CPC Party Secretary of Chongqing in late 2007, as one Chongqing legal consultant and planner noted in an interview, Chongqing officials were notified of plans to develop a new economic zone through phone calls from senior bureaucrats in January 2008 (Interview, Chongqing, March 2012; author's translation). This strongly suggests that Bo had in mind the idea of producing Liangjiang New Area even prior to his appointment (more on this shortly). Soon thereafter, the municipal government invited a research team from the central government to explore how to advance the '314' agenda.

A visiting team of 219 centrally-approved experts arrived in June 2008. Following the visit, the Chongqing government proposed 12 new policy suggestions to the Chinese State Council, the primary objective of which was arguably to secure central government financial support to develop Liangjiang New Area as part of a broader 'national strategy'. The visiting research team added 7 additional suggestions to the State Council, which led to a '12 + 7' policy framework. These proposals were subsequently approved in an elaborate, 38-item work plan known as the 'No. 3 document'. Pledging to enhance Chongqing's 'strategic position' (*zhanlüe diwei* 战略地位) and, in an indication of the enduring significance of the 'whole nation as a chessboard' philosophy (ref. Chapter 1), calling for Chongqing policymakers to plan in relation to the national 'whole situation' (*quanju* 全局), the No. 3 document established the parameters for re-designating $1200\,km^2$ of land north of the core urban area into a territory of national priority. This territory is Liangjiang New Area (see Figure 6.1).

As Table 6.2 indicates, this most recent round of spatial reconfiguration within the municipal scale occurred in tandem with the '314' integrated developmental

Table 6.2   The changing developmental trajectory of Chongqing since its repositioning in 1997.

Year	Development	Characteristics
1997	Granted province-level city (*zhixiashi*) status	• First (and at the time of writing, still the only) city in China's interior to be granted this status • Discursively identified as the 'window' to and 'dragon head' of the west • Entrusted by the central government to fulfil '4 big things' (*sijiandashi*), namely (i) the resettlement of migrants affected by the Three Gorges Dam construction; (ii) the reformation of old industrial bases; (iii) the 'lifting' of the poor in rural villages; and (iv) the protection of the biophysical environment.
2000	Official implementation of the 'Great Western Opening Up' (*xibudakaifa*) cross-provincial program	• Chongqing aligned to the experimentation of 'big city lead big village' (*dachengshi dai danongcun*) developmental approach • Identified as a geographic hub that connects the central and western parts of China
2007	'314' Overall Strategy identified (*zongtibushu*)	• Launched by China President Hu Jintao during the national 'Two Meetings' in March, in tandem with Chongqing Party Secretary Wang Yang • The strategy would form the guiding framework for subsequent socioeconomic reforms in the municipality
2009	'No. 3' document (*guofa sanhao wenjian*) issued on Chongqing's socioeconomic reforms	• Issued January; termed 'A number of opinions regarding the enhancement of Chongqing city's urban–rural integration reforms and development'; marked the implementation of the '314' integrated developmental strategy • Also reflected the beginning of attempts by the Chongqing government to 'scale up' economic development
2010	Designated a 'national central city' in 'National Urban System Plan' (*quanguo chengzhentixi guihua*)	• Plan devised by Ministry of Housing and Urban–Rural Development; announced in February. • Only city amongst the five designated cities – Beijing, Tianjin, Shanghai and Guangzhou – not located along the coast
2010	Liangjiang New Area officially opened for development	• Unveiled on 18 June the same date when Chongqing became a province-level city • An economic development zone spanning 1200 km², within which experimental policies are devised and implemented • Positioned as 'one door, two centres and three bases' (*yimenhu, liangzhongxin, sanjidi*), namely (i) the main entry point to western China; (ii) both an import–export trading centre and a financial centre in the Yangtze upstream region; and (iii) the bases for modern manufacturing, export-processing and the commercialisation of R&D outcomes

(*Continued*)

Table 6.2 (Continued)

Year	Development	Characteristics
2010	Two major socio-spatial policies launched	• Public rental housing construction begins in February • *Hukou* conversion program officially implemented in (after a series of pilot experimentations)
2013	Administrative autonomy mooted for Liangjiang New Area	• On 30 November 2013, the Chongqing announced a series of decisions to detailing how administrative autonomy is to be granted to Liangjiang New Area; allows the New Area greater flexibility to respond to national and global economic opportunities • Indicates a continuity of the spatial project launched by the Bo Xilai-Huang Qifan administration; whether social reforms will be expanded remains unclear
2016	Liangjiang New Area consolidates local and national position	• Autonomous role of the Liangjiang New Area regulatory commission further clarified in June 2016 • Liangjiang New Area to become the 'focal zone' (*hexinqu* 核心区) of Chongqing's pilot free trade zone (重庆自由贸易试验区) • Chongqing continues double digit GDP growth amidst national-level growth slowdown; Liangjiang New Area contributes around 13% of Chongqing GDP

Source: *Chongqing Ribao* (8 October 2007; 14 January 2012a); *People's Daily Online* (10 February 2010). Data compiled and translated by author.

strategy (in March 2007) and the decision to confer 'national central city' status on Chongqing (in February 2010). It was a multi-scalar process to enhance Chongqing's competitiveness in relation to other places. Working on the basis of what was introduced in Chapters 1 and 3 as the 'move first, experiment first' approach, Bo Xilai and Huang Qifan, the-then Chongqing mayor, proposed to the central government a list of policy advances across different regulatory domains. Central to their proposal was the institution of the country's third 'nationally strategic' economic development zone (the other two at the time were Pudong in Shanghai and Binhai in Tianjin; see Figure 6.1).

The 'promotion' of Liangjiang New Area to 'nationally strategic' status was made possible by an integrated strategy to effect localised supply-chain integration; to attract targeted transnational corporations (TNCs) to relocate their offshore operations to Chongqing; and to connect Chongqing to major logistical hubs in the European Union (EU) through a borderless rail link, an unprecedented move that effectively cuts transcontinental transport time to under two weeks. In this regard, Liangjiang New Area symbolised the Bo government's ambition to transcend a core limitation commonly associated with interior regions – poor access to transport routes. As Huang Qifan remarked, reducing transaction costs associated with transport was a major goal in the spatial reconfiguration process:

> The largest constituent of foreign capital investments in China involves export processing, this comprises 50% of China's annual US$2,500 billion worth of exports. This export-processing model has been active along the coast for 20 years, employing almost 200 million workers. The special feature of this model is 'two heads beyond', i.e. the original spare parts come from beyond, and the retail markets are located beyond, while only the processing base and assembly plants are located along China's coasts. If export-processing in Chongqing is done the same way, the spare parts would be shipped from overseas to China's coasts, then transported 2000 km to Chongqing, the transport costs and flow time would mean this model is unworkable. There has not been export-processing in inland China because goods transport has been an issue. (Huang, interview with *Chongqing Ribao*, 20 May 2011; author's translation)

Ironically, Chongqing's emergent 'pivotal position' in contemporary China could be attributed to its prior lack of integration within the national and global economy. This was an outcome generated by the 'ladder step' developmental approach first introduced in Chapter 1. Specifically, this approach reconfigured economic heterogeneity across Chinese territory: rather than differentiated by function within a nationally-insulated political economy, some places (e.g. the coastal seaboard) became more connected to the global economy than others. Variegated developmental pathways thus ensued and re-defined uneven development across the country (ref. Chapters 1 and 3). This unevenness simultaneously contained the seeds for fresh development in China's western interior. In a bid to 'catch up' with the industrialising coastal city-regions, the Chongqing government actively sought to develop the infrastructural foundations by undertaking the financial risks. In other words, the inequality generated by the 'ladder step' approach was a crucial precondition for new layers of space-shrinking investments. Put another way, the ability of the Chongqing government to invest in infrastructural construction laid the groundwork for the subsequent '314' developmental program and the strategy to 'scale up' the designation of Liangjiang New Area (ref. Tables 6.1 and 6.2 and the following section). Yet, Huang's interpretation that 'market signals were not in place' does not mean the state automatically had to intervene (ref. Huang's interview with *ENN Weekly*, 21 January 2014). As Cui Jian, the former chairman of State-owned Assets Supervision and Administration Commission in Chongqing (Chongqing SASAC), the 'market' was simply unwilling to drive infrastructural development in Chongqing following its re-designation into a centrally-governed municipality because of the 'ladder step' approach:

> In fact the 'eight major investments' policy was premised on conditions in which market mechanisms were weak; the government thus organised SOEs into a model to finance the expansion of economic developmental capacities. To take the

formation of the first investment sector, the public water works group, as an example, a major battle took place. Around 2000, when the Three Gorges Dam needed to store water, the central government decided to establish 16 wastewater management plants along the Yangtze River. Through national bonds, each firm that was selected to build and manage the plants would receive 50% of the project fee, at the time the national bond bureau stated that this amount would go to whoever takes on the project. (Cui, interview with *21st Century Business Herald*, 9 March 2013b; author's translation)

As Cui's comment shows, the CPC was originally keen to allow private firms autonomy in the construction and management of the plants, but this did not materialise because domestic banks (the majority of which were and remains state-owned) were plainly unused to dealing with private-sector led development in the western interior:

The 16 private bosses [who bid] agreed happily, and 50% of the project fee was thrown in. When the central government came to check in 2002, it discovered not a single of those bosses used the money, all offered the same reason: banks would not offer additional loans [to make up the other 50%], they did not believe wastewater management plants could generate the ability to repay the loans. The National Development and Reform Commission (NDRC) added another 25% of the fees, following which the plants were constructed but were not put in operation. The private bosses stated the financing costs of operating the plants were not planned holistically, even though the construction fees were paid, who would pay the maintenance fees? The management of these plants became an irresolvable conundrum. (Cui, interview with *21st Century Business Herald*, 9 March 2013b; author's translation)

In turn, the Chongqing government had to implore the NDRC to delegate more autonomy to run its infrastructural construction. This intervention reinforced the lack of private market strength in Chongqing and arguably re-constituted direct state involvement in the contemporary Chinese political economy:

The Chongqing municipal government submitted a report to the NDRC stating that the original model of financing only the construction of a project was incorrect. It might as well allow the Chongqing government to construct a large-scale waterworks management corporation that could undertake the national bonds, fixed assets and projects associated with the 16 wastewater plants and engage in centralised financial management. (Cui, interview with *21st Century Business Herald*, 9 March 2013b; author's translation)

Huang and Cui's comments reveal a particular condition specific to China's developmental trajectory: the 'national market' was produced and held together by a strong state before it was 'liberalised' selectively in 1978. Underpinning this strong state were Mao-era policies that fortified central control of the national socio-economy and Deng to experiment with the 'ladder step' approach. Viewed in relation to the analytical framework presented in Chapter 3, it was inherited institutional strength, repurposed through coastal-oriented industrialisation projects in the 1980s and 1990s, that entrenched and reproduced the interventionist regime in Chongqing. And, going by Huang's argument, there is no implicit necessity for it to 'retreat' vis-à-vis private producers once the state is embedded in the process of capital accumulation:

> Why must there be a retreat? How would this retreat be done? It should not be done on the basis of idealism, on the feeling that the government should not control enterprises and should retreat during good times. However, in reality we are retreating on the basis of market principles. (Huang, interview with *China Reform*, November 2010; author's translation)

One distinct phenomenon of this entwinement with 'market principles' is the mode of 'retreat' – state control of its corporations was delegated to and subjected to the functions of capital markets. Apart from offshore listings, Chongqing-based SOEs have been encouraged to raise capital and/or invest overseas as part of the national 'Go Abroad' developmental strategy launched after the 2008 global financial crisis. This cross-border market involvement by state-linked economic entities consequently gives a new definition to state-capital relations: the state could affect and respond strategically to the 'rational' calculations of transnational circulatory capital in the name of social security. Cui Jian, the former Chongqing SASAC Director, highlighted the rationale behind this proactivity:

> What we are pushing forth in Chongqing is an integration of 'going out' and 'bringing back,' which leads to a model that 'kills three birds with one stone'. Through overseas mergers and acquisitions or sole financing of investments, we actualise 'going out'; through overseas mechanisms, we raise capital and re-invest in Chongqing, facilitating 'bringing back'; through the reinvestments we enjoy the manifold effects of security and long-term economic development. (Cui, interview with *The Economic Observer*, 28 November 2011; author's translation)

The transposition of the provincial-level state into a transnational economic actor investing in longer term 'security' and 'long-term economic development' complicates the state-centric accounts of economic development that are presented in Chapter 1. Contrary to the approach in Guangdong, which sought to enable state rescaling through attracting investments from centrally-owned SOEs and firms based in Hong Kong and Macau, the Chongqing case saw the state directly involved as *both* regulator and participant. This two-pronged involvement – a distinct legacy of Mao-era governance – was justified as a form of necessary paternalism. Cui Jian, the former Chongqing SASAC director, elaborates on this point:

> I think the objective of local SOEs is to play a leading role in the local economy, unlike central-administered state-owned enterprises, which seek to become world leaders. Through its SOEs, Chongqing created the 'third finance' [state-owned assets financing public projects under a cost-recovery model], which generates tax and dividends, helps the government with investment and reduces fiscal expenditures, thereby enhancing local government's urban–rural development, and makes possible the transformation of public finance. I think this ought to be the focus of local SOEs. (Cui, interview with *The Economic Observer*, 28 November 2011; author's translation)

Wen Tiejun, a prominent China-based economist, sums up the effect of this path-dependence on post-2007 socioeconomic reforms in Chongqing:

> The advantage in Chongqing is the ability to concentrate energies and launch huge projects (*jizhong liliang gan dashi* 集中力量干大事)...Because the developmental capital in Chongqing is too high, most ordinary private firms would be ineffective on this front; moreover, because the developmental parameters are to a large extent outcomes of the discussions between the Chongqing government and the central government, Chongqing received 12 beneficial policies. As such, the developmental capital could only be taken on by sizeable state-owned enterprises. (Wen 2011: 65; author's translation)

Taken together, these comments point to the need to establish how the so-called 'Chongqing experience' of development is at once a product of and a process to circumvent the legacies of experimental policies introduced in the first two decades of post-Mao reforms (cf. Figure 2.1, Chapter 2). Viewed in relation to these conditions that made possible socioeconomic reforms in post-2007 Chongqing, the equation of socioeconomic reforms in the municipality with Maoist resurgence appears problematic. Specifically, national-level reforms generated the conditions of possibility for the Chongqing government to spearhead China's economic restructuring after the global financial crisis struck in late 2008 (more on the constitutive effects of the crisis shortly). It is in this sense that the emerging phenomena in Chongqing are not expressions of Maoist path-dependency – they appear to take on Maoist logics of socioeconomic regulation *because of* policies launched in the post-Mao era.

One defining characteristic of the apparent path-dependency was the prioritisation of public ownership – in particular land, but now including other means of production like public utilities, finance and info-communications – in the Chongqing economy. This was (and remains) a characteristic inherited from the Mao-era and which was not as well-reformed relative to those in the coastal city-regions. In an intriguing twist, this inherited characteristic undergirded the massive investments in infrastructural and social projects that subsequently enabled the Bo Xilai government to launch the Liangjiang New Area project. Rather than negate public ownership, the Chongqing government began to actively restructure the SOEs under its charge after Huang Qifan took over as mayor in 2002. The initial goal, introduced previously as *badatou*, was to enable the state to drive investments in eight major infrastructure-related sectors. To the Chongqing government, state involvement in China's relatively under-developed regions was an economic necessity at the time. Speaking in 2014 with ENN Weekly, Huang explains the *badatou* project was launched because of weak market signals. It was for this reason that the project would not 'have longevity, its scale and characteristics are changing according to time and circumstances. These firms would have to withdraw after completing its construction task within a specific domain, they either shut down or change jobs to become a concrete industry.'

The Chongqing government's decision to intervene ahead of markets constituted *the production of space by time*. Its ability to take the initiative in integrated infrastructural development not be uniquely attributed to a conjuncture (in the context of this study, the period after Bo Xilai took over as Party Secretary of Chongqing): inherited institutions both enabled and encumbered Bo's ability to drive change. As this section has demonstrated, the roots of the post-2007 change extend at least to the 'Third Front' construction drive in 1964. In turn, it raises further questions on how Chinese state power was reproduced through its adoption of market-based governance.[4] It is thus imperative, as mentioned earlier in Chapters 1 and 2, to make clear the connections between the latest series of socioeconomic reforms and the constitutive conditions that made these reforms possible.

Liangjiang New Area as a Post-crisis Regulatory 'Fix'

State rescaling, as Chapter 3 has argued, is a contextually-specific process of reconfiguring regulatory relations. A new round of experimental policies was set in motion in Chongqing after the global financial crisis deepened in 2008. Interestingly, as discussed earlier in Chapter 4, this corresponded with the Guangdong government's response to the crisis. The remaining part of this chapter will demonstrate how the national designation of Liangjiang New Area was a multi-scalar process made possible by political strategies of 'scaling up' by actors based at the provincial scale, changing external economic conditions and macro-level adjustment considerations (or *hongguan tiaokong*, as discussed in Chapter 2) in the central government.

Just as the new Bo Xilai administration was tasked to implement the '314' approach, the global financial crisis in 2008 opened up new possibilities to justify the production of a new economic growth pole. To the-then Chongqing mayor,

Huang Qifan, the global financial crisis provided an 'opportunity' for the formation of a new economic development zone in the western interior:

> In the June–July period of 2008, we organised a whole meeting of the municipal committee and decided on developing a liberal growth pole in the interior. Actually in February and March of 2008 we had already begun preliminary discussions on a strategy to push forth economic liberalisation. The opportunity for this was the global financial crisis, there was a global [economic] contraction of around 30%. (Huang, interview with *21st Century Business Herald*, 17 April 2010; author's translation)

Weng Jieming, the former Director General of the management committee of Chongqing Liangjiang New Area, offers an expanded elaboration of the contextual factors that made possible the designation of a 'nationally strategic' zone in Chongqing:

> In my exchanges with friends, all would raise these questions: Why is Liangjiang New Area situated in Chongqing? Why did the national government choose to establish Liangjiang at this time? I feel it is related to the broader context of national development. Two criteria are extremely important. One, the macro-level strategy of national economic development has reached a time when major decisions about adjustments must be made. Two, the launch of the new decade of the 'Great Western Opening Up' program required important action. These two criteria are directly related to the need to establish a 'carrier' (*zaiti* 载体) in the hub on China's interior. (Weng, interview with *China Web*, 27 September 2010; author's translation)

As Weng's comment shows, another major experimental project of significance is the cross-provincial 'Great Western Opening Up' development agenda. Mooted in 1999 by the Jiang Zemin administration and launched a year later, the agenda has been largely placed at the discursive level, with no targeted growth poles or systematic restructuring strategies launched at the metropolitan scales. As introduced in Chapter 1, the primary contribution of these broad regional programs is to establish a geographical consciousness. Because more targeted strategies were not immediately forthcoming in the early 2000s, the Chongqing government decided to be proactive in driving economic development in the municipality.

In itself this observation underscores the fact that the overarching developmental agenda of the Hu Jintao government could not move away from a path based on coastal bias. More crucially, it demonstrated the Hu government did not have a targeted approach to overcome this bias prior to the global financial crisis. In this regard, Weng's comment was telling: Liangjiang New Area became reality because of both the global financial crisis and the consequent attempt to effect change at the subnational levels to ensure nationwide structural coherence (ref. framework presented in Chapter 3):

Right from the beginning Liangjiang 'carried on' the role as the breakthrough point of China's strategic readjustment strategy. After the global financial crisis, the export-dependent, external demand-dependent developmental model that drove the reforms and liberalisation for the past few decades was challenged. Under this situation, we wondered what methods could enable high-speed economic development, and determined it should be driven by domestic demand. This lead role has been on the agenda for many years, and the establishment of Liangjiang New Area, in tandem with the development of the broader interior hinterland, would realise this. Second, regional development would require a growth pole, and within the growth pole there must be a nucleus…In this new decade of Great Western Development, there is a need to balance the east and the interior, and bridge the flow from south to north. The western cities of Chongqing, Chengdu and Xi'an could all become regional growth poles for the interior, but there must be a nucleus, hence the leaders in the State Council mentioned many times that [because] Chongqing is a 'national central city' of China, a nucleus could be formed within it to push forth regional development in the west and in turn reduce uneven development. These two reasons constitute the broad background for the formation of Liangjiang New Area. (Weng, interview with *China Web*, 27 September 2010; author's translation)

On their own, the interaction of an external phenomenon (the financial crisis), a domestic spatial project (the Great Western Development Program) and Chongqing's designation as a 'national central city' could not explain why other cities like Chengdu and Xi'an were not preferred as 'nationally strategic' sites. Indeed, while the Chinese central government was under pressure to ease the impacts of the global financial crisis on the Chinese economy by setting in motion an economic restructuring plan, it was simultaneously under pressure from many provincial and municipal governments across the country to be 'scaled up' as key focal points for the national restructuring strategy. As Chapters 4 and 5 have shown, the Guangdong government was proactively seeking to 'scale up' three key cities while it launched a 'double relocation' restructuring project. And as presented in Chapter 3, the PBoC governor, Zhou Xiaochuan, acknowledged that fierce competition exists at the inter-provincial level when it comes to securing the authority to launch financial reforms.

Against these bottom–up initiatives to gain 'move first, experiment first' power, sensitivity to policies and practices instituted before the 2008 global financial crisis is necessary. New information further reinforces the previous point that the Chinese central government did not have a coherent growth-pole strategy for the western interior prior to the 2007 financial crisis. As Xiao Jincheng, director of the Institute of Spatial Planning and Regional Economy of the NDRC, revealed to the *21st Century Business Herald*, the CPC never intended to launch new 'nationally strategic' growth poles for the entire state spatiality after 2006:

A 'nationally strategic new area' is first and foremost a 'hat' – it is to be the focal point from all directions…Thinking back, the state did not have an objective plan to designate a third or fourth 'new area'. On the contrary, after demarcating Tianjin

Bianhai New Area, the central government had the intention to put a full stop [to this strategy]. Subsequently Wuhan, Shenyang and Chengdu were all fighting to get the growth pole status but was disregarded by the state, no 'hat' was conferred on any party. After approving the demarcation of Binhai New Area, no other new area [at the sub-provincial scale] was approved until 2010, when Liangjiang New Area was designated. (Xiao, interview with *21st Century Business Herald*, 27 December 2013b; author's translation)

Viewed in relation to Weng's comment, Xiao's recollection raises an important question: was the designation of 'national central city' really the deciding factor why Chongqing was chosen as the site to launch a 'growth pole nucleus'? As their comments and the empirical analysis in Chapters 4 and 5 demonstrate, the central government was not short of alternatives when it came to selecting new 'nationally strategic' new areas. Crucially, from Weng's comments in two separate instances, it can be inferred that the ability of the Chongqing government to change its positioning 'from the "third front" to the "first front" was an outcome of its "proactivity" '. First, there was significant political will within Chongqing to engage with transnational circulatory capital:

Chongqing will transform from the 'third front' to the 'first front', to become the frontier of liberalisation in the [Chinese] interior. Confronting the global financial crisis and the huge transformations related to China's national developmental readjustment, the Chongqing municipal committee and government proactively included itself in the national strategy just before the new Great Western Development plan was launched. It grasped the initiative to enable Liangjiang New Area to become the launch-off project, the motor and important window, of this new plan. (Weng, interview with *Chongqing Ribao*, 6 August 2010a; author's translation)

Perhaps more importantly, such proactivity was precisely what the central government *expected* of subnational governments:

As for why the central government preferred Chongqing, this has to do with the proactive working initiatives in the Chongqing government. As early as 2008 the municipal committee has established the opening up of the interior as a working target, a plan to establish an experimental zone for the interior was developed back then. Since this zone lies between the Yangtze and Jialing rivers, hence it took on the name of 'Liangjiang' ('Two Rivers'). Subsequently we built on the adjustments in national developmental strategy and the launch of the new decade of 'Great Western Opening Up' and continuously lobbied ideas to the central government, a central–local coordinated movement was established at this unique conjuncture, all parties came to a consensus, and Liangjiang New Area was born. (Weng, interview with *China Web*, 27 September 2010; author's translation)

In another interview with *Chongqing Ribao* in 2010, Weng pointedly stated that the rescaling of Liangjiang New Area was not 'an innocent and simple coincidence':

From this year [2010], the central government has intensified the launch of the Great Western Development Program. In May the Political Bureau of the CPC Central Committee...indicated it would prioritise opening up the western region in the overall national-level regional coordination strategy...Similarly in May, the State Council issued [On the decision to approve the establishment of Liangjiang New Area] (National document 36, 2010). At an important conjuncture in which the state pushed through a new round of Great Western Opening Up, Liangjiang New Area emerged, this should not be an innocent and simple coincidence. (Weng, interview with *Chongqing Ribao*, 7 September 2010; author's translation)

Through two interviews with Chongqing-based academics who participated in several planning seminars prior to the establishment of Liangjiang New Area, the Chongqing government led by Bo Xilai was instrumental in converting Liangjiang from idea to reality by building on the broader backdrop of the '314' integrated program.

The path for restructuring was already set for Bo by the earlier government led by Wang Yang. It was originally only reforms on urban–rural integration, and Bo did a great job publicising these reforms. It was only after he came that the Liangjiang project was launched, it was built on the social reforms set in place for him to carry out. (Interview with academic A, Chongqing, March 2012)

Just before Bo Xilai was appointed in Chongqing, the Ministry of Commerce [where Bo was Minister] signed a memorandum with the Chongqing government [led by Wang Yang, who was posted to Guangdong] to explore developing Chongqing into a more open economy. Immediately after his transfer, he inspected the area where Liangjiang is located, and in January 2008 Chongqing officials received instructions to think about how to develop a new economic zone. In April 2008, less than five months after Bo's Chongqing appointment, the Chongqing government proposed to then Chinese Premier Wen Jiabao 12 new policies for the development of a 'liberal' area. Wen immediately agreed to send people from the central government to conduct more research. Two months later they [the Chinese State Council] dispatched 219 people to launch a series of studies. (Interview with academic B, Chongqing, March 2012)

Taken together, the comments of Weng, Xiao and the two academics strongly suggest the post-2007 Chongqing government adroitly built on the agenda set by his predecessor Wang Yang (in tandem with the central government). Academic B's comment further reinforces the notion that Bo already fashioned the idea of a growth pole nucleus prior to his posting to Chongqing. Yet pushing through his proposed ideas did not mean pleasing only the 'experts' from the central government; the Bo administration, as an interview with a planner in Beijing (February 2012) reveals, encountered strong opposition from economic geographers with strong ties to the central government. One of these geographers, X from a highly renowned institution in Beijing, was famous in policymaking and academic circles for a geographically-functionalistic framework that portrayed coastal China as enjoying intrinsic comparative advantage in trade over interior China. X was of the opinion that establishing Liangjiang New Area was a waste of resources because transportation costs would be too high, and sought to have his opinion heard by the State Council. This duly delayed the approval process. While it was not publicised, the planner indicated that the-then Chongqing mayor, Huang Qifan, travelled to Beijing to allay academic X's concerns (interview, February 2012). How this was done was unclear, but the State Council eventually passed the proposal in January 2009.

Viewed in relation to the production of Hengqin and Qianhai New Areas (see Chapters 4 and 5), these findings lend further support to the argument that the Chinese central government did not have a spatially-targeted rescaling agenda; the emergence of experimental policies in the 'nationally strategic new areas' were simultaneously opportunistic reactions by the central government to macro-level challenges and 'bottom up' initiatives by provincial-level governments to gain an advantage in what Li and Wu (2012: 55) term 'excessive' territorial competition (ref. comments by Zhou Xiaochuan in Chapter 3). Xiao Jincheng of the NDRC sums up the *spontaneity* of the state rescaling process:

> There was no logic to abide, at most it was to actualise regional coordination. In practice it was to place the new area against a broader backdrop of integrated experimental reforms. For instance, after integrated reforms were introduced in Pudong and Binhai, new reforms on efficient resource use were introduced in Wuhan, Changsha [in central China] and urban–rural integration reforms introduced in Chongqing and Chengdu. (Xiao, interview with *21st Century Business Herald*, 27 December 2013b; author's translation)

The Chongqing officials not only lobbied the central government for 'move first, experiment first' policies, they were actively courting TNCs. Differing from other industrialising projects in China, where a common strategy was to begin first with the redrawing of spatial boundaries before attracting firms with some preferential policies, a concerted strategy was drawn up to bring in firms before the boundaries of Liangjiang New Area were demarcated. As Section 6.2 has explained, a major factor that enhanced Chongqing's competitiveness was its ability to enhance time–space compression. Connectivity to global markets was deemed crucial to the success of a growth-pole nucleus in the western interior. For this reason, the Chongqing government invested heavily in infrastructural facilities in order to enhance the municipality's economic–geographical position.[5] Once the requisite infrastructure was in place, transport costs were no longer a limiting factor. State officials then went on a proactive drive to enrol transnational economic actors. The Chongqing mayor at the time, Huang Qifan, offers an elaborate explanation of this enrolment process:

To break through this predicament, we fully grasped the fortuity presented by reconfiguration of the global IT industries and made two innovations in Chongqing on the export-processing model for inland regions. First we built a full-flow supply chain for parts and final product (*quanliucheng canyelian* 全流程 产业链), transforming the horizontal division of labour into vertical integration [of the production process], with 70–80% of spare parts to be produced locally. This transforms the 'two heads beyond' model into 'retail beyond, spare parts supply and integration within' ['one head within, one head beyond'], which greatly reduces or even removes the transport costs associated with spare parts supply. (Huang, interview with *DushiKuaibao*, 7 March 2011; author's translation)

This new model moves beyond the quest for just-in-time (JIT) production, which technically could occur even within global production networks, to an almost total territorialisation of these networks in Chongqing. Just as some firms seek cost advantages by choosing to produce in-house, the Chongqing government was seeking cost advantages by bringing together all levels of specific production networks like computing notebook manufacturing *in situ*. And these levels include some aspects of advanced producer services:

We changed a pattern of the offshore location of settlement services for high value-added production. We discovered this problem, which was very common but nobody seemed to give it much attention, when we were trying to bring in Hewlett-Packard. We brought in Hewlett-Packard's US$100 billion worth of settlement accounts when it relocated its Asia-Pacific settlement centre to Chongqing from Singapore, because of this, the taxes drawn from these services and the banking service fees would be retained in Chongqing. Through these innovations, we successfully constructed a '2+6+200' notebook production industrial agglomeration, comprising Hewlett-Packard and Acer as two 'dragon heads' [i.e. lead firms], Foxconn, Quanta, Inventec, Wistron, Pegatron and Compal as six major subcontractors, and 200 odd spare parts producers. This develops a 100 million-capacity, US$100 billion value computer notebook production base, the largest in Asia, which significantly transforms Chongqing's economic structure. (Huang, interview with *DushiKuaibao*, 7 March 2011; author's translation)

The Chongqing government's ability to effect the relocation of Hewlett-Packard's supply chain did not happen in a geographical–historical vacuum. As mentioned earlier, the basic infrastructure was already in place; what was needed to make Liangjiang New Area a success is its ability to generate income and employment. What ensued was a delicate balancing of place-specific institutional power with corporate power that eventually enhanced the attractiveness of Chongqing as a frontier of experimental reforms vis-à-vis other locations in the western interior. By extension, it reinforces the point in Chapter 3 that state rescaling is in part an outcome of subnational developmental agendas. The Chongqing government understood the global financial crisis was causing Hewlett-Packard to rethink its spatial division of labour, and decided to offer new conditions to the firm in the hope that it would generate a positive response:

> When Hewlett-Packard was planning production increases at the beginning of 2009, most sales of computing products were affected by the global financial crisis and began to dip, only the sales of notebooks continued to increase because of technological shifts, total output was expected to double to 320 million by 2012. Where would this production increase be done, through the Chinese coastal region or somewhere else? Hewlett-Packard was searching for a place to expand production capacity, if they relied on their old strategy they could continue their 'two heads outwards' approach...But then the labour and land costs were increasing significantly in Shanghai, at that moment if there was a place that had all sorts of costs, including that of labour, that was lower than Shanghai, then Hewlett-Packard would be moved.
>
> At the beginning of 2009 we approached Hewlett-Packard for discussions. At the time they felt our transport costs were high, they wanted us to subsidise it. We then proposed our 'one head out, one head in' model, making the point that our transport costs could only be lower than those along the coast. And we told Hewlett-Packard at the time, if you only intend to produce several tens of thousands of units in Chongqing, spare part suppliers would not come over to Chongqing. Yet if Hewlett-Packard would place an order of 40 million units, then the suppliers would also come, and an agglomeration effect would emerge. (Huang, interview with *Economic Information Daily*, 21 April 2010; author's translation)

After Hewlett-Packard agreed to the Chongqing government's proposal, plans were swiftly implemented to form a place-based 'temporary coalition' involving actors originating from different locations. This came in the form of driving extensive supply-chain migration in the transnational notebook-manufacturing sector:

> Immediately after speaking to Hewlett-Packard, I went to Taiwan on 9 February 2009 to visit Terry Gou [chairman and founder of parts giant Foxconn]. Within three minutes of entering his conference room I went direct with him, saying I was not there to implore him to attract investments but to do a transaction: Hewlett-Packard would like to produce 40 million units of notebook computers in Chongqing because we wanted to have integrated production. You produce spare parts in the coastal region for 50 million units of computers, but never produced a full unit. If I give you 15 million of the 40 million units Hewlett-Packard intends to produce, Foxconn could make full units in Chongqing. But I request you bring along the whole system of spare part production from the coastal region to Chongqing, because the costs of making spare parts are lower than in the coastal region. Terry Gou and I hit it off right away. (Huang, interview with *Economic Information Daily*, 21 April 2010; author's translation)

At this point, it would be useful to note that the proactive engagement with transnational circulatory capital did not in itself guarantee state rescaling or policy experimentation through Chongqing. Occurring more than one and a half years before the official unveiling of Liangjiang New Area, the proactive approach to embed transnational capital undergirded the actualisation of the '862' industrialisation strategy (ref. Table 6.2). Targeting transnational lead firms like Hewlett-Packard raised the profile of the municipality and proved that it was possible for a location in interior China to offer new productive capacities for global capital. While it was unclear if special subsidies were given to Foxconn or other incoming original equipment manufacturers, there was a parallel effort to subsidise costs of labour through the public rental housing construction and the conferment of basic social benefits for migrant workers (see Chapter 7). This arguably helped to keep already lower labour costs lower than it would be if firms must factor in costs of social reproduction (e.g. building dormitory complexes).

The proactive attempt to bring in foreign capital generated sustained economic success. 54 Fortune 500 firms announced they would establish operations in Liangjiang New Area when it was launched; 118 of these firms were established by the end of 2013 and generated export value of US$30.5 billion, or 44% of the Chongqing total (*Xinhua*, 22 January 2014b). By June 2018, 147 Fortune 500 firms were operating in Liangjiang New Area and were central in tripling the territory's GDP since 2009 (*Chongqing Shibao*, 22 June 2018). This growth was arguably not envisaged when the '314' developmental approach was launched in 2007. As Huang further explains, he had to fend off competition from other western Chinese cities for TNCs and suppliers, further affirming the argument that the establishment of a growth pole nucleus in Chongqing was not predetermined by the Chinese central government:

> Chengdu tried to compete with us for Inventec and Foxconn. At the beginning they just followed the usual approach of attracting capital, namely through some preferential policies and the low cost of labour as attractions. They did not understand the Taiwanese are just suppliers for foreign firms, they will go where these firms go. I would not avoid telling our brothers in the western interior, the crux of attracting capital is not to attract suppliers, it is to attract Acer, Dell, Hewlett-Packard. And yet these [lead firms] are fierce competitors in reality, hence when Chongqing has Hewlett-Packard, I would not try to bring in Hewlett-Packard's nearest rival Acer, but I will bring in a 5th-ranked competitor such as Toshiba, because they do not pose as big a threat to Hewlett-Packard and could coexist peacefully. (*Xinhua*, 22 January 2014b)

From the dynamic interaction between the Chongqing government's initiatives and the post-2007 shifts in national developmental strategies, the logic of state rescaling becomes clear. As Chapter 3 has argued, rescaling tendencies emerge not through the central planning agenda; it is dependent on the interaction between different levels of the state system and new opportunities to shape the (re) allocation of capital. For the Chongqing government, a particular opportunity opened up after the global financial crisis went full-blown in 2008. As firms contemplated reorganising their spatial divisions of labour to exploit cost-competitiveness in the post-crisis era, the challenge for the Chinese central government was to encourage these firms to remain in China. Chapter 4 has shown how firms in Guangdong came under significant pressure after the global financial crisis; the same pressure was evident in other industrialised city-regions along the coastal seaboard like Shanghai, Kunshan, and Tianjin. The difference between these city-regions in China is that the Chongqing government, just like the Guangdong government, used the crisis as the foil to 'scale up' the significance of targeted territories. In Guangdong these were Hengqin, Qianhai, and Nansha New Areas; in Chongqing it was Liangjiang New Area.

With an aggressive approach to attract transnational capital, the establishment of Liangjiang New Area provided a very timely geographical opportunity for the Chinese central government to confront the global financial crisis without destabilising its ability to keep China attractive to transnational circulatory capital. As such, it could be argued that economic–geographical reconfigurations that support the basic structure of the Chinese state and its objectives will be privileged. Clearly recognising this, the Chongqing government was active behind the scenes to push for 'nationally strategic' institutional change. This ultimately culminated in the plan for Liangjiang New Area, as laid out in Table 6.3.

Conclusion

This chapter has argued that 'nationally strategic' reforms instituted in Chongqing after 2007 were not predetermined by the Chinese central government. Through re-contextualising these reforms against a longer chain of events originating and occurring in different scales, the chapter shows the transformation of Chongqing to be at once conditioned by inherited policies extending back to the Mao era and a new experimental agenda to place Chongqing on the world map of high-end manufacturing. As such, while the anti-Mafia crackdown may have brought into question the importance of the rule of law; while the 'red song' singing campaigns may have made the post-2007 reforms seem ideologically motivated; and while the move to reduce social inequality proved popular (see next chapter), it would be simplistic to construe these reforms as signifying the re-emergence of the Cultural Revolution (ref. introductory section, this chapter).

The 'nationally strategic' reforms to be launched in Chongqing came to be because of specific conditions that were never predetermined by the CPC. This chapter has identified these conditions as, namely, (i) Chongqing's historical role as an industrial base; (ii) Deng Xiaoping's coastal bias in the initial round of reforms in the post-Mao era; (iii) the centrally-mandated Three Gorges Dam construction and the corresponding re-designation of Chongqing into a broader municipality governed by the central government; (iv) the reconfiguration of state involvement in the Chongqing economy since the early 2000s; and (v) the '314'

Table 6.3   Key aspects and implications of newly promulgated plan for socioeconomic development in Chongqing Liangjiang New Area during the 12th Five-Year Plan, launched in September 2011.

Aspects	Corresponding references in plan	Implications
Three integrated 'impulses' (*sanda dongli*)	• Integrated approach to urban–rural development; 'open up' the western interior; and endogenous innovations • To be actualised through (i) bringing in targeted, large-scale manufacturing and service industries; (ii) staggered urban functions to enhance overall municipal capacities; (iii) improvements in public service provision standards and people's livelihoods; and (iv) the establishment of export process zones and zones of foreign capital concentration.	• The first targeted spatial strategy in China that projects 'outwards' to other scales through the basis of urban–rural integration • Accentuates the social basis of economic development • Builds on the '314' integrated development strategy
Spatialising production networks	• Referred as '862' approach to economic development, namely • '8' manufacturing industries (Automotive, infocommunications, rail and transport, high-end assembly, aerospace, recycling, new energy, biomedicine) • '6' service industries (Integrated logistics, commerce, international conventions, cloud computing, corporate HQ and service outsourcing) • '2' hubs (finance and innovation).	• Similar to 'picking winners' strategy of East Asian developmental states • Spatial strategy precedes economic reality: suggests the constitutive importance of supply-side targeting • Plans to become internationally-renown offshore financial centre highly similar to reform objectives in the GPRD (see Chapter 5)
Key regulatory emphases	• To achieve enhanced development through breakthroughs in reforms of the urban–rural dual structure • To continue pushing for the central government to approve policy experiments in 'key domains' (*zhongdian lingyu*)	• Strong involvement of the central government suggests Liangjiang New Area constitutes an aspect of 'central localism' (ref. Chapters 1 and 3)

Source: Plan from www.people.com.cn (2013); author's translation and analysis.

integrated developmental strategy launched by Wang Yang, in tandem with Hu Jintao. As the analysis has shown, the interaction of these spatial projects and inherited policies more accurately illustrates why the Chongqing government intervenes significantly in the municipal economy.

To be sure, the ability of the state to transform economic geographies in Chongqing since 1997 could be attributed to an inherited state–state collaborative legacy of the Mao-era: state-owned banks simply found SOEs more credit-worthy than private economic actors, in turn impeding the involvement of the non-state private sector in the foundational sectors of the Chongqing economy. Once successfully reformed, however, it was hard for private economic actors to rival these SOEs. This effect of first-wave reforms in the post-Mao era explains why the Chongqing government could – and, more crucially, continues to believe should – be the primary actor that drives economic–geographical restructuring (ref. preceding section). In this respect, the 'big' state in Chongqing is as much an outcome of post-Mao reforms that privileged coastal city-regions as it is about policies implemented in the Mao era.

Viewed in relation to the spatial strategies implemented at the national scale, this lack of faith by non-state investors could be construed as a direct outcome of Deng's ladder-step approach to capital allocation (see Chapter 1). The lagging economic growth in the western interior simply provided few opportunities for private investors to appropriate surpluses. Without discernible revenue streams accruing from FDI-inflows and a clear developmental strategy, state-owned banks inevitably preferred working with state-backed partners. The success of Deng's 1978 spatial strategy – another indicator of institutional strength – thus paradoxically provided the platform for the consolidation of the Chongqing government as an economic actor. In other words, it was instituted economic–geographical disadvantage that impelled the Chongqing government to reproduce its strong grasp on the municipal economy. Ironically, the expansion of market mechanisms across China not only did not encumber state intervention, it enhanced the state's economic strength (ref. preceding section).

What this interaction of successive policies demonstrates, then, was the tensions associated with establishing new developmental pathways and the conditioning effects of earlier policies (ref. analytical framework developed in Chapter 3). While the Bo Xilai government wanted to 'liberalise' (kaifang) the Chongqing economy and subject it to continuous market reforms, it had to do so on the paths determined by the '314' strategy, namely to drive economic growth on the basis of urban–rural integration. The '314' strategy was in itself a response to the Mao-era 'dual structure' and the growing inter-regional inequality caused by Deng's spatial strategies (ref. Chapters 1 and 2). This explains why the developmental agenda for Liangjiang was driven by 'three integrated impulses' (ref. Table 6.2). As such, it would be more accurate to view the 'Chongqing model' as a simultaneous attempt to fulfil the agenda to effect spatial egalitarianism (drawn up by Wang Yang and Hu Jintao) and the unprecedented quest for a 'nationally strategic' growth-pole for capital accumulation in the western interior (drawn up by Bo Xilai and Huang Qifan). Placed in relation to the attempt to reproduce economic competitiveness in Guangdong through new experimental policies (ref. Chapters 4 and 5), it could further be argued that state rescaling in China is highly uneven: each targeted location possesses its unique historical legacy, and the outcomes of policy experimentation would necessarily bear the effects of inherited institutions. Would it be plausible, then, to extend experiments in one location to the national scale?

While the combination of these two experimental agendas may have been unforeseen, it inevitably raised uncomfortable questions on China's developmental approaches. As Chapter 7 will elaborate, socially-progressive reforms were presented as ideologically complementary to the economic liberalisation process, an issue the previous regimes of Deng Xiaoping and Jiang Zemin circumvented. It could thus be argued that the rescaling of Liangjiang New Area enabled the social reforms mandated in the '314' developmental strategy to take on a higher profile; these reforms, which would have occurred even without the production of Liangjiang New Area, became 'nationally strategic' by association. And because the reforms could unsettle entrenched practices elsewhere in the country that gave rise to the Chinese economic growth 'miracle', strong opposition concomitantly emerged.

Endnotes

1   Whilst officially designated a city, the term 'city-region' is used to describe Chongqing for two reasons. One, it occupies a huge land mass of 82,400 km², more than double the size of Taiwan; second, a significant portion of this land mass is still farmland and, until July 2012, less than half of the official 29 million population are registered as 'urban' residents.
2   As an entity, the 'Republic of China' is no longer recognised by the United Nations. It continues to exist in name, however, with its base in Taiwan Island, and continues to be recognised internationally by a handful of countries. For a full geo-historical discussion, see Lim (2012).
3   In March 1985, the Chinese State Council mooted a plan to rescale a 'Three Gorges Province' (*sanxia sheng* 三峡省). This plan would encompass the current Chongqing administrative area as well as western Hubei province. However, for reasons still unknown, the plan was officially cancelled in May 1986.
4   While it appears like a paradox, in reality it is a variant of neoliberal transformation, in which the quest for a 'free market' is undergirded by a strong state.
5   The initial financing channel was, like many local governments, through debt-financing by major banks linked to the Chinese central government. Currently there are more financing channels (public listing in Hong Kong being a favoured mode; the other is to plough back revenue into new investments), although debt-financing remains at around 50%. (Author's participation in academic seminar on 'Chongqing model', March 2012.)

Chapter Seven
State Rescaling in and Through Chongqing II
The Politics of Path-dependency

Introduction

> *The urban–rural dual structure is the primary obstacle restraining the integration of urban–rural development.*
>
> Communiqué of the 3rd Plenum of the 18th Congress of the Communist
> Party of China (CPC) (Xinhua, 12 November 2013b)

The proclamation 'Development is the absolute principle' (*fazhan shi yingdaoli* 发展是硬道理) by China's then-leader, Deng Xiaoping, in the early 1990s arguably triggered 'GDP-ism' as the guiding developmental ideology. In the subsequent decade, 'development' was narrowly interpreted as the inexorable quest to maximise surplus value capture through the market exchange of commodities. The inevitable consequence was pronounced socio-spatial inequality. It was against this national context of economic–geographical unevenness and rising social angst that led Wang Shaoguang (2002: n.p.) to proclaim 'social equity is also the absolute principle'. After Hu Jintao became China's leader in 2003 and launched the ideology 'Scientific Perspective on Development' (*kexue fazhan guan* 科学发展观), the quest to reconcile high-speed economic expansion with more equitable redistribution gained momentum. As Hu would later add, 'development is the absolute principle, [but] stability is the absolute task, without stability, nothing can be done, whatever we have achieved will also be lost' (www.chinanews.com, 1 July 2011;

On Shifting Foundations: State Rescaling, Policy Experimentation and Economic Restructuring in Post-1949 China,
First Edition. Kean Fan Lim.
© 2019 Royal Geographical Society (with the Institute of British Geographers).
Published 2019 by John Wiley & Sons Ltd.

author's translation). In this regard, Hu's '314' injunction for experimental reforms in the Chongqing city-region arguably marked a spatial shift in the CPC's developmental ideology (ref. Table 6.1, Chapter 6). The 'absolute' task of producing social stability in China, as the socioeconomic reforms in Chongqing indicate, is directly premised on addressing the consequences of instituted uneven development.[1]

This focus on more balanced developmental pathway provided the political basis for fortifying the social safety net in the municipality. As the then-Party Secretary of Chongqing, Bo Xilai, explains, 'amongst our country's four province-level cities, Chongqing's rural area and population is higher than those of Beijing and Tianjin combined, if we do not do a good job at urban–rural integration, the core city area [of Chongqing] may advance significantly, but the areas to the south and north would stagnate, urban–rural disparity would then be widened…Chongqing must become a big platform for [policy] experiments' (interview with *Economy and Nation Weekly*, 10 January 2011; author's translation). To Yang Qingyu, Head of Chongqing's Development and Reform Commission, the reform objectives in Chongqing reflect a nascent but necessary shift in the 'feeling for stones' philosophy first introduced in Chapter 1:

> Crossing the river by feeling the stones is done in areas of shallow water, after reaching today's deep water areas, the river can't be crossed by feeling the stones, [we] will get drowned, there's a need to form a bridge, but bridge formation requires integrated techniques. Hence back then [i.e. the nascent years of reforms] it's possible for a lone soldier to make a breakthrough [across the river], but this cannot be done today. Yet everybody also cannot move forward together, this is the complicated nature of [contemporary] reforms. (Yang, interview with *China Business Times*, 8 March 2012; author's translation)

The implication of Bo and Yang's comments is clear: the state apparatus, rather than private market actors, must proactively ensure the developmental process allows everyone to eventually 'cross the river'. This discursive representation distinguished the reforms in Chongqing from those of/in other provinces. What the Chongqing government purported to mitigate were the social costs and spatial consequences associated with urban-based industrialisation, an integrated approach that was unprecedented in the post-Mao era. Bo puts the Chongqing government's path-changing experimental approach in elaborate perspective:

> The core value system of socialism does not only exist in philosophy and spirit but in developmental thought and pathways. To insist on socialistic common affluence, in my personal understanding, is a fundamental condition of the socialist value system…While we encourage competition and acknowledge the differences in the incomes of social residents, at the same time we place high importance on social equity, and how the middle- and lower-income group live their lives. *Egalitarian redistribution and high-speed economic growth can be attained simultaneously*; Chongqing has been taking this path these past few years. At the same time as pursuing social

fairness and justice, we achieved high and quality growth, I feel it is possible, both can be done simultaneously. (Bo, open interview with media, transcribed by *Xinjingbao*, 6 March 2012; author's translation, emphases added)

Bo's reference to the 'core value system of socialism' was arguably the primary reason why critics were so quick to paint the Chongqing reforms as a radical, left-leaning strategy to reproduce the Cultural Revolution. If anything, the Bo government's emphasis on the simultaneous enhancement of social and economic life aroused memories of the integrated developmental approach in the Mao-era People's Communes (ref. Chapter 2 on the role of the communes). As Chapter 6 has shown, however, Bo's primary contribution during his tenure as Party Secretary of Chongqing was arguably the demarcation of Liangjiang New Area as the first growth pole nucleus in the western interior. Rather, it was Wang Yang, Bo's predecessor and the espouser of 'market values' in Guangdong (ref. Chapter 4), who instituted the reforms to 'place high importance on social equity'. In sequential terms, then, the Bo government was adding a strong economic layer to the '314' policy while putting in practice the social reforms initiated by Wang and approved by Hu Jintao.

Extending the line of argument presented in Chapter 6, this chapter argues that the experimental social reforms in post-2007 Chongqing represent a concerted attempt to (re)generate a socially-equitable developmental pathway last instituted during the first few years of Chinese state formation. These reforms are (i) embedded across a longer chain of events that stretches back to the Mao era as well as (ii) entwined with state rescaling processes initiated by the Bo Xilai government. As discussed in Chapter 2, the generation of GDP became increasingly privileged during the post-Mao era. State rollback of social benefits followed consequently, first with the termination of the People's Communes in 1984 and then the fiscal system overhaul in 1994. In colloquial parlance, the politico-economic system increasingly 'determined heroes by GDP figures' (*yi GDP lun yingxiong*, 以GDP论英雄), and it was this logic that drove experimental reforms in Guangdong (ref. Chapter 5). At one level, the Bo government was arguably following the same formula of economic development by launching large-scale industrialisation in Liangjiang New Area; what differed was it also had to carry out social reforms that the Hu Jintao regime identified as inevitable in March 2007. The emergent reforms in Chongqing were thus a *geographically-targeted* re-introduction of an integrated developmental approach previously sidelined by Deng Xiaoping and his successors.

Specifically, the centrepiece of the experimental reforms in Chongqing was to remove an intractable Mao-era institution – the urban–rural 'dual structure' – that undergirded post-Mao economic–geographical development (cf. Chan 1994; Chan and Buckingham 2008; ref. Chapter 2). This chapter will analyse and evaluate the rationale of two interrelated social reforms – the conversion of rural migrants into urban residents and the provision of public rental housing for these new urban migrants – in relation to their critiques. Through this analysis, it can be further affirmed that the socioeconomic reforms in Chongqing are constituted by a dynamic interaction of policies instituted over an extended period – they are neither an expression of a definite national developmental agenda nor a unique 'bottom–up' initiative (ref. the problematisation of the centralisation–decentralisation binary in Chapter 2).

There are three parts to this chapter. The next section discusses the rationale of the new social reforms, paying specific attention to how these reforms are framed discursively as complementary to the drive for expanded industrialisation in Chongqing. The third section will evaluate the critiques of the reforms. While these critiques portrayed the reforms as an expression of political populism in Chongqing, they take on a different shade when contextualised against a broader chain of events. It will also be shown that the Chongqing government financed the reforms on a cost-recovery model, which further discredits the claims that the reforms in Chongqing signify a return to Mao-era economic governance. Rather, the section makes the argument that the reforms mark an intriguing use of market mechanisms to fulfil state developmental goals. The concluding section will discuss how the post-2007 socioeconomic reforms in Chongqing reinforces the notion, introduced in Chapter 2, that market-oriented reforms have enhanced the capacities of the Chinese party-state to plan and allocate resources.

Experimenting with Spatial Egalitarianism: Public Rental Housing Provision for Migrant Workers

Against a national context of escalating housing prices, Chongqing's then-Party Secretary Bo Xilai made a sudden announcement in December 2009: in the coming decade, his government would launch a series of large-scale public rental housing projects to provide affordable housing to low-to-middle income Chongqing residents (estimated to be 30–40% of the total population). If unease was the initial reaction for (nominally-) private developers,[2] surprise would probably be next: the construction of the symbolically-named *Minxin Jiayuan* (Good Home of People's Hearts 民心佳园) – the first project of public rental flats – began at the end of February 2010. The timing of Bo's announcement, which targeted housing a third[3] of the population in Chongqing's urban centre, appeared highly strategic in retrospect: it dovetailed with the Chinese State Council's January 2010 injunction to tackle the nationwide housing price inflation mentioned previously.

The primary reason why the large-scale public rental housing construction generated strong attention was arguably because it underscored a fundamental problem associated with market-like rule in China. Low-income residents in the cities, many of whom are migrants, have been excluded from the housing market following the commodification of housing in 1998.[4] Social discontent grew concomitantly, exacerbated by the massive inflation[5] in property prices in the decade that followed housing commodification. Youqin Huang, a prominent analyst of developments in China's national housing policies, explains:

> [H]ousing prices have skyrocketed in cities, with the national average housing price increasing by 250% in the decade between 2000 and 2010. The housing price-income ratio classifies much of China as 'severely unaffordable' in terms of housing. In big cities like Beijing and Shanghai, a modest apartment can cost multiple millions of yuan to purchase, and thousands of yuan to rent, making housing affordability the top concern of most low- and middle-income households. As a result, millions of migrants have been completely left out of the 'Chinese dream', with few owning homes in cities and most living in extremely crowded, poor quality dwellings. (Huang, writing for *The Diplomat*, 14 May 2013, n.p.)

Of particular significance, Huang adds, is the connection between housing unaffordability and one's *hukou* status, which in turn accentuates the tension between housing commodification, a post-Mao development, and the *hukou* institution, an enduring product of the 'socialist high tide' of 1958:

> Migrants are generally more vulnerable in the housing market due to their lower incomes and the discriminatory Household Registration System, or *hukou* system (often called an internal passport system), under which migrants are not considered 'legal' residents in cities despite living and working in there over the long-term. Without local urban *hukou*, migrants are not entitled to welfare benefits such as subsidised housing. Even in Shenzhen, the city of migrants, local *hukou* is required to access low-income housing. In others cities like Beijing, several years of local *hukou* is required before applying for low-income housing. (Huang, writing for *The Diplomat*, 14 May 2013, n.p.)

Because residents holding urban *hukou* almost always enjoy better social benefits than those holding rural *hukou*, rural residents often find themselves disenfranchised and discriminated against when they relocate to cities. While the abolition of this institutionalised inequality is regularly touted, its structure remains intact and is seemingly intractable (ref discussion in Chapter 2). In Kam Wing Chan's (1994) evocative terms, these migrants continue to live in 'cities with invisible walls'. These seemingly intractable issues pertaining to the provision of low-income housing in the city to non-urban *hukou* residents would be addressed in Chongqing.

In contrast to the Mao-era 'dual structure', the public rental housing project is an attempt to open up the public housing sector – previously the preserve of selected state-owned enterprise (SOE) employees – to the city's lower-income residents (from Chongqing's rural districts as well as other provinces) who might otherwise decide against moving into or near the Chongqing core urban area where major industrial zones (particularly Liangjiang New Area, the flagship industrialisation project of the Chongqing government) are located. As mentioned in Chapter 6, state subsidisation of social reproduction is a crucial ingredient for the fulfilment of nationally-mandated industrialisation objectives in and through the city-region. Huang Qifan, the former Chongqing mayor who implemented these reforms, summarises this state-firm relationship:

> The large scale construction of public rental housing not only guarantees weaker groups the ability to lead life with dignity, it will also improve Chongqing's investment environment and stimulate the local economy. Through reducing the cost of living, [affordable housing] can attract talents and firms and infuse life into Chongqing's economic development; this is way more meaningful than land-based financing [tudicaizheng 土地财政] through high property prices. (Huang, interview with CCTV; transcribed by *Chongqing Wanbao*, 5 September 2010; author's translation)
>
> Because of the separation between the government and [private] firms, the government and society take over [responsibility for] the workers' lives; firms need only build factories, they need not build dormitories. Hence they could reduce their investments; this is beneficial for attracting investors and capital. (Huang, interview with *China Reform*, November 2010; author's translation)

In addition, Huang and Tang Zongwei, the deputy director of the Liangjiang New Area Administrative Committee, elaborate on how the large-scale state investments in public rental housing cannot be defined narrowly as process to reduce social discontent. Rather, and in an approach that diametrically contrasts Wang Yang's subsequent 'double relocation' project in Guangdong (ref. Chapter 4), investments in social reproduction were deemed to be integral to the national strategy to establish Chongqing – and Liangjiang New Area in particular – as the economic fulcrum (*jingji shuniu* 经济枢纽) of interior China:

> The entry of rural workers into the city cannot lead to new poverty enclaves, nor can they be allowed to buy houses; they can't afford it. Since there are several million people waiting to enter the city [i.e. Chongqing], we must find a way to solve their housing needs...so long as you are a worker in Chongqing, so long as you have worked for three, five years, the city needs you to work, then you can apply for rental housing. For new university graduates who enter Chongqing, so long as a firm employs you, then you can apply for rental housing. In sum, our housing safeguard system runs on the basis of urban–rural integration. (Huang, interview with www.news.cn, 8 March 2012; author's translation)
>
> Just within Liangjiang New Area we aim to construct around 8.5 million m², this does not include the public rental housing planned elsewhere in Chongqing municipality...The meaning of our public rental housing project is distinct from conventional notions, our project is to assist those who are just about to commence employment or become entrepreneurs. These groups of people come from everywhere, they either do not have housing or temporarily cannot afford housing, if you don't assist, they can't settle down. Hence our public rental housing is targeted at a million university graduates, researchers and high-quality industrial workers, it is to

help them enter Liangjiang, to help them integrate in this city, and it simultaneously resolves the problem of industrial support. (Tang Zongwei, interview with *21st Century Herald*, 27 November 2013b; author's translation)

Within the broader reform package, the rental housing project's socioeconomic objective complements another urban–rural integration policy, namely the flexible conversion of rural migrant workers' *hukou* into 'urban' status (see Table 7.1). As Youqin Huang's previously-mentioned comments indicate, most migrant workers

Table 7.1   Key characteristics of evolving *hukou* reforms in Chongqing.

Geo-historical context	Objectives of *hukou* reforms in Chongqing	Policy shifts
• Instituted in 1958 by the Mao Zedong government to control population movement • Citizens broadly categorised as 'urban' and 'rural'; *hukou* (household registration affiliation) determined social benefit provision, administered by provincial governments; • Primary aim was to control the right to the city by preventing mass rural-to-urban migration; hence widely viewed as a form of socio-spatial apartheid and institutionalised inequality • Right of movement significantly relaxed since the 1990s as industrialisation expanded in cities, but the system remains relatively intact	• To guarantee the interests of those who work in the city without urban residency • To improve rural productivity through the transference of spare labour to the urban areas • To improve the urban population structure and alleviate the pressure of an aging society • To expand domestic demand • To speed up the process of urbanisation.	• Reforms officially launched on 15 August 2010 with the introduction of 'The Opinion on *hukou* Institutional Reforms in the Urban–rural Integration of Chongqing Municipality' • Based on a 'voluntary, compensatory' (*ziyuan, youchang*) principle • Expanded provision of social benefits for urban *hukou* holders: partially financed through fiscal redistribution, with firms and urban residents contributing the rest • New urban migrants will enjoy, amongst other social benefits, access to public rental housing if they meet the stipulated conditions • Switching the right to use land is not a precondition of *hukou* conversion[a]; the transference of land rights is voluntary and all transfers will be compensated

Sources: Chan and Buckingham (2008); *Caijing* (30 August 2010a; 30 August 2010b); data compiled and translated by author. Cross-ref. discussion of *hukou* institution in Chapter 2.

[a] After this standpoint is cross-referred with the published regulations, the relationship between *hukou* conversion and ownership of land use rights remains unclear: the new regulation states plainly rural residents have three years to relinquish their land if they choose to convert *hukou*, which contradicts the discursive statements that *hukou* conversion is not entwined with land rights transfer. The disentanglement of *hukou* conversion with land rights transfers – which precludes the creation of a 'third category' of citizens with urban *hukou* status *and* rural land use rights – therefore remains a policy 'grey zone'.

nationwide could not qualify for subsidised public housing in cities across China. The reform of *hukou* institutions in Chongqing is thus a spatially-targeted attempt to overturn the socio-spatial exclusions engendered by this enduring Mao-era legacy. Crucially, as Chapter 6 has explained, this reform would have taken place independent of the rescaling of Liangjiang New Area; it was set out by the '314' developmental agenda instituted by then Chongqing Party Secretary Wang Yang. Central to this agenda, as the-then Chongqing mayor, Huang Qifan, was at pains to point out, is a recognition of peasant migrants' inherent right to be treated equally as full Chinese citizens, not just as economic units of labour power:

> Assuming a city has 3 million registered residents and 7 million peasant workers, after a few decades, when the city does not need these rural workers, they are dismissed back home, pushing the population down to only the [original] 3 million registered residents. This urbanisation process is unhealthy, abnormal, it violates international conventions, it violates market principles, it violates the principles of fairness and justice. (Huang, interview with *China Economy and Informatization*, 18 January 2012; author's translation)

> Improving the treatment of peasant migrant workers is a matter of human rights; it is a problem of citizens' rights. To treat peasant migrant workers benevolently is a matter of our conscience; it is a matter of governors' consciences...Over the past few years Chongqing arranged for 3.6 million peasant workers to convert their *hukou*. The crux of our program is to facilitate peasant workers to enter the city, not to have more peasants in the city. With the development of urban-based industries, many new industrial and service sector positions have become available; they need to be filled by a large number of peasant workers. In the several years during which these workers help fill the open job positions, it would be unreasonable if you did not provide them with urban *hukou*, it would be unjust, this would be half-baked urbanisation, not a true urbanisation. (Huang, interview with *China Economic Weekly*, 18 March 2013; author's translation)

To the-then mayor Huang, the impossibility of launching instant reforms is precisely why the *hukou* institutional reforms must be implemented in stages. After all, socio-spatial transformation – which is akin to forming a bridge for a large party to cross a wide and deep river – requires time:

> The urbanisation process cannot envisage a situation where, after two groups of people were kept apart for thirty or forty years, several hundred millions of rural workers are accumulated in cities and then suddenly it is announced their hukou will be changed. This [hukou conversion] should not be a sudden occurrence, it should begin now; it should be a natural and continuous transformation process. (Huang, interview with *China Economy and Informatization*, 18 January 2012; author's translation)

Huang's emphases on 'fairness', 'justice' and 'citizens' rights' in Chongqing signify an attempt to actualise Deng Xiaoping's vision of inter-regional equality as presented in Chapter 1. This attempt could not be construed as a destabilising outcome of decentralisation, however. While the large-scale *hukou* conversion of peasants into 'non-agricultural' citizens and, concomitantly, the provision of social services to accommodate their relocation to urban areas are integral aspects of a subnational developmental agenda: they are at once enabled by centrally-facilitated experimental power (i.e. *xianxing xianshi quan* 先行先试权; ref. analytical framework in Chapter 2). Indeed, it was only after China's President Hu Jintao issued a '314' integrated development strategy for Chongqing in 2007 that the city-region's policymakers immediately went about launching policies to reduce the inherent rural–urban barriers (*chengxiang bilei* 城乡壁垒; ref. Table 6.1, Chapter 6). The 'local' attempt to enhance citizenship rights in Chongqing is thus more aptly conceptualised as a centrally-driven path generating effect.

One primary expression of this effect is the provision of public rental housing to rural residents expected to convert their *hukou* to the 'non-agricultural' (or, more simply, 'urban') category. Expanding housing supply to support *hukou* reforms (50% of rental housing stock is allocated for new peasant migrants) is an experimental attempt to combat house price inflation, which as mentioned earlier is a phenomenon triggered by the nationwide commodification of housing provision. This process takes on social significance as spiralling housing prices would certainly have the greatest impact on newly-minted urban residents who, like their predecessors in many of China's urban-industrial parks, are most likely to be employed in low-wage, labour intensive industries or service sectors in Chongqing's urbanised industrial areas (particularly in the 'nationally strategic' Liangjiang New Area).

The integrated approach to reform the 'dual structure' and produce a new growth pole nucleus in Liangjiang New Area led to a surge in both GDP and fixed capital formation (see Figure 6.1, Chapter 6). Until the appointment of Bo Xilai as Party Secretary in late 2007, the annual GDP expansion rate ranged between 8 and 12%. This rate quickly moved beyond 20% within three years. Chongqing's GDP registered a 21.4% expansion in 2010; its 26.3% growth in 2011 topped China's provincial-level GDP growth charts (Chongqing Municipal Bureau of Statistics 2013). Perhaps more intriguing is the scale of its fixed capital formation, which surged from 56% of GDP in 2007 to 88% of GDP in 2011, almost double the (already-high) national average. This positive relationship between high GDP growth and high fixed capital formation strongly suggests the latter constitutes a significant portion of GDP. With SOEs, collectives and joint ventures driving at least 40% of these investments since 2008,[6] it lends support to the argument that the rapid GDP growth in Chongqing was an outcome of a 'big' state reminiscent of the Mao era. While these robust growth figures seemingly support Bo's argument that high speed GDP growth could occur in tandem with the enhancement of social equity, they ironically became the symbol of what is wrong about the Chongqing 'model'.

Market-based Maoism?

Oppositional voices from national academic and media circles grew concomitantly with the implementation of socioeconomic reforms in Chongqing. Specifically, the Chongqing reforms triggered a nationwide debate on 'making a bigger cake' (*zuoda dangao* 做大蛋糕, or the economistic developmental approach widely associated with that adopted in Guangdong) in relation to 'sharing the cake well' (*fenhao dangao* 分好蛋糕, or redistributing existing gains). This debate illustrates two seemingly incommensurable logics of socioeconomic regulation across China: to some, redistributing resources to enhance socio-spatial justice was the antithesis of expanded capital accumulation, while others viewed the enhanced stability accruing from centrally-approved redistribution as the precondition of expanded accumulation. Viewed in relation to the dynamic analytical framework presented in Chapter 3, this debate is not narrowly confined to the Chongqing reforms – it more accurately exemplifies instituted uneven development by the Mao administration.

After Deng launched experimental reforms in 1978, the Chinese central government's approach to national governance suggests clearly that 'cake sharing' is less important than 'cake making'. 'To compete for foreign investment', Wu (2009: 842) observes, 'local states [in China] maintain a stable social order and invest heavily in productive infrastructure such as roads and airports, while being reluctant to roll out welfare services'. Where and when redistributive policies are implemented, they tend to take on a distinct bias towards producing facilities for economic production. By implication, then, only areas considered strategic to GDP-growth are deemed worthy of fiscal support. It was arguably for this reason that the huge costs of public rental housing construction and *hukou* conversion in Chongqing became the primary focal point of prominent critics. How, indeed, could expanding access to social benefits contribute simultaneously to growing the cake of capital?

Speaking to *Investor Journal*, Mao Yushi, the Chairman of the Unirule Institute of Economics in China and winner of the Cato Institute's 2012 Milton Friedman Prize for Advancing Liberty, believes the Chongqing government is overextending its finances through debt-financing in order to generate political populism:

> The construction of rental flats is an attempt by the government to curry favour with the people, but where is the government's money from? From taxpayers. [...] Without the creation of wealth, there is no wealth to redistribute; is Chongqing more efficient than other places in wealth creation? The strong development of state-owned enterprises in Chongqing…is a way to decrease efficiency in productivity. Under such a situation, where do you find so much money to do good things for people's lives? It's very possible that [the policies] depend on bank credit.
> (Interview with *Investor Journal*, 26 March 2012; author's translation)

Wu Jinglian, a highly-influential economist and a senior economic advisor to China's State Council, echoes Mao's views:

> Such populism means sparing no expense to please the population, but it caused many problems for Chongqing's budget. One cannot invest such huge resources while having little regard for where such resources come from. One also has to consider whether such investments produce returns. (*South China Morning Post*, 26 March 2012)

These comments exemplify a belief in the inherent superiority of a self-regulatory market. As Wu Jinglian has argued elsewhere, only two possible developmental trajectories exist for the Chinese economy in the current conjuncture: one involves the 'perfection' of market reforms, with political power circumscribed by the rule of law, the other a 'cul-de-sac' characterised by state capitalism and crony capitalism (*China Entrepreneur*, 10 September 2013). For Mao Yushi, the market is quite simply 'the most successful institution' in the course of human history (Interview with *Shenzhen Special Zone Daily*, 4 June 2012). Placed in relation to concrete empirical facts in China, however, this market fundamentalism is shaky in two ways.

First, the argument that the socioeconomic reforms in Chongqing were untenable just because they reflect populist politics needs unpacking. In concrete terms, the pressing needs for alleviating what was (and remains) a growing rich–poor divide in China's largest municipality – which in itself is an outcome of the coastal- and urban-bias in earlier developmental strategies – would have been sufficient to merit state (pre-emptive and ameliorative) action (cf. Chapter 6). Interestingly, Wu Jinglian showed he has never forgotten class struggle with an observation that 'socioeconomic contradictions [in China] have already reached a critical point, we cannot continue to blunder, a revolution would happen if we continue to blunder' (*China Entrepreneur*, 1 November 2013; author's translation). Mao Yushi has similarly acknowledged that 'the biggest drawback of markets is the creation of rich-poor inequality. Markets create this problem, hence it cannot be resolved by markets. Its correction has to come from extra-market forces' (Interview with *Shenzhen Special Zone Daily*, 4 June 2012). Such is the serious nature of social problems in contemporary China, Mao Yushi (2013) went on to illustrate at book-length the colossal worries of the national populace since the 1978 'liberalisation' reforms.

These observations raise an intriguing question for the conceptualisation of 'politically popular' socioeconomic reforms in Chongqing: if the need to ease social tensions is so pressing; if the people are already intensely worried by the reforms brought about by deepening market-like rule; and if market mechanisms cannot ease these tensions, would experimental programs to ameliorate these tensions constitute a 'blunder' on the part of the CPC? Or would these programs constitute attempts at producing more socially benign forms of regulation?

China-based scholar Qiu Feng offers an interesting counterpoint on the interconnections between political populism (*mincui zhuyi* 民粹主义), 'crony capitalism', and the Chongqing reforms:

> I very much dislike those who criticise [on the basis of] populism. We can see on Weibo [China's version of Twitter] people from Chongqing's political circles who claim China's biggest danger in future is populism. I wrote a commentary on this: what lies before our eyes is crony capitalism, what we now need to consider is how to solve this problem. Populism is a reaction to crony capitalism, there is first crony capitalism before there is populism. Now many intellectuals follow the wave to oppose populism, they equate the 'Chongqing model' to populism. I feel there should not be such a definition, resolving problems associated with people's livelihoods are separate from populism. (Qiu, part of academic discussion at the Unirule Institute of Economics, Beijing, 6 December 2012. Transcript published in *China Review*, 15 January 2013, n.p.; author's translation)

The second problematic assumption is associated with the approach to resolve problems associated with people's livelihoods, namely the involvement of credit markets in the redistributive process in Chongqing. At one level, this concern about potential budgetary delinquency is warranted: many local governments in China have indeed encountered difficulties repaying debt to (nominally-) private creditors. It could indeed be argued that lax regulation and monitoring of lending practices to government-linked institutions is an ingredient that feeds crony capitalism. This situation is well summed up by Larry Lang, a prominent Hong Kong-based analyst of Chinese economic development:

> Honestly speaking, I am really worried about the local governmental debts in China. In 2009, local governments could still borrow from new sources to repay old creditors; by the end of 2012, what they did was blatantly request extensions of repayment deadlines…I do not dare to imagine what measures our local governments and banking system will adopt to address wave after wave of maturing debts. (Larry Lang, 19 December 2013, n.p.; author's translation)

It should be clear, however, that the cause for worry is not about whether state agencies should borrow from credit markets – it is about the ability to repay. In this regard, Mao Yushi's point that the debt-independent ability to create wealth (presumably a task exclusive to private firms) should undergird fiscal redistribution becomes a paradoxical assumption: in a capitalistic system, wealth creation is driven by debt. As Chapter 4 has shown, state rescaling and policy-experimentation in Guangdong was also strongly contingent on debt-financing (and this was never raised as an issue by influential commentators like Mao Yushi and Wu Jinglian). Mao Yushi ostensibly assumes that only private firms – rather than state-owned enterprises – should undertake debt because of their presupposed 'efficiency' at wealth creation, while the redistributive process (which, in China, includes financing state-owned enterprises) should not imbricate wealth creation processes (which theoretically means any activity that qualifies for debt-financing). The public scholar Qiu Feng offers a different interpretation on the debt-financing issue:

> Of course, many people criticise these explorations in Chongqing, they claim that the Chongqing government has undertaken a lot of debt from banks; that its finances is already bankrupt. To be honest, this kind of argument is strange because almost all local governments are shouldering heavy debts, most local governments debt-finance their 'projects for face-creation' (*mianzi gongcheng* 面子工程) or 'projects for political results' (*zhengji gongcheng* 政绩工程). To just criticise Chongqing based on this point seems unreasonable. What can be investigated is how much has been done [in Chongqing] to improve people's lives. (Qiu, part of academic discussion at the Unirule Institute of Economics, Beijing, 6 December 2012. Transcript published in *China Review*, 15 January 2013, n.p.; author's translation)

To gain a stronger understanding of the experimental reforms unfolding in Chongqing, Qiu Feng calls attention to the objectives of redistributive policies. The redistributive process, for Qiu, cannot be assumed to be negative ab initio; it is important to evaluate whether redistributive policies can enhance socioeconomic development. This chapter concurs with Qiu's contention. As Section 7.2 has shown, the housing project is entwined with the *hukou* conversion and industrialisation drives. These drives are in turn outcomes of a path-changing developmental agenda developed by the former Chongqing Party Secretary Wang Yang and Chinese President Hu Jintao (ref. '314' program introduced in the previous chapter). Against this broader contextual backdrop, the redistributive reforms were at once necessary and contingent: they were necessary because the Chinese central government mandated the time has arrived to tackle fundamental contradictions associated with the 'dual structure' instituted in 1958; they were contingent because there was no guarantee that the experimental reforms would succeed and, in turn, be expanded across the country.

To facilitate a more incisive evaluation, it would be useful to consider in greater detail the logics underpinning the provision of public rental housing. If the Chongqing government had chosen not to 'invest' – the term Wu Jinglian uses to describe fiscal redistribution – in the expansion of housing stock (rental or ownership-based), the investment risk would first be borne by real estate corporations, before being ultimately transposed onto individual buyers through a quasi-cornering process (current low-income urban residents and new migrants need somewhere to stay after all, which translates into definite demand for housing). Yet the Chongqing policymakers determined the lower income strata of Chongqing's urban migrants would find privately-supplied apartments unaffordable (see introductory section, this chapter). This meant the lower income strata could not generate effective demand for privately-supplied housing unless high-risk credit provisions were given.

Viewed this way, the Chongqing government's involvement as a housing stock supplier is a means to reduce social risks of housing debt defaults by lower income earners à la the US sub-prime mortgage crisis of 2007–2008. Specifically, it illuminates the negative relationship between excessive housing debt and the suppression of effective consumer demand. As it has become clear in many cities within China and even across East Asia, when housing supply is controlled by private property groups (usually an oligopoly), many individuals would take up – and be locked-in by – housing debts that extend very long terms (30–35-year repayment periods are commonplace).Whether individuals – especially new rural migrants – in Chongqing are able to satisfy their loan obligations over two to three decades is moot per se (the US national sub-prime housing market crash is a case-in-point); the more crucial question, rather, is the impact of long-term property debt servicing on economic expansion: when individuals have to devote significant portions of their disposable income to repay housing loans, their ability to purchase consumer goods naturally decreases. Given the inverse correlation between high housing debt and effective consumer demand, the public rental housing project in Chongqing offers a potential economic 'return', to use Wu's parlance. It frees up disposable income, which raises the propensity for direct consumption and/or an increase in savings (and which then fuels new rounds of investments). This thus fulfils a key developmental objective – the expansion of domestic demand – of China's 12th 5-Year Plan (2011–2015).

To address the critiques of these 'investments' directly, it would be more pertinent to assess the feasibility of the repayment channels. There were also differences on this point. Writing for the *Financial Times*, private investor Liu Haiying offers this interpretation of Chongqing's budgetary position towards the end of Bo's tenure as Party Secretary:

> From an analysis of [Chongqing's] fiscal revenues...the GDP for 2011 was 1 trillion *yuan*, and fiscal revenue (including extra-budgetary collection) was 290.8 billion *yuan*...this constitutes 29% of GDP. That fiscal revenue takes up such a high proportion represents a strong extraction of social resources by the government.Yet such a strong extraction cannot satisfy its expenditure demands: in 2011 Chongqing's fiscal expenditures hit 396.1 billion *yuan*, almost 40% of GDP; the national proportion was 23%. From this calculation, the nominal fiscal deficit is 105.3 billion *yuan*, 10.5% of GDP, far above the 3% international warning levels. (*Financial Times*, 9 March 2012, Chinese edition; author's translation)

Speaking to South Korean media,Wu Jinglian went further by pronouncing the end of the 'Chongqing model' based on a lacerating interpretation of Chongqing's budgetary and fixed capital investment statistics of the same conjuncture:

> Chongqing's GDP [for 2011] was around one trillion yuan, and fiscal revenue was around 150 billion yuan [sic]. Yet within the last year [2011], the fixed assets

investments in Chongqing reached 750 billion yuan, benefiting from this, Chongqing's economic growth rate also reached 15% [sic], how long can this situation be sustained? The 'Chongqing model' has failed and is finished. (Wu, interview with *Korea JoonGang Daily*, 17 April 2012; author's translation)

However, Liu and Wu's presumptions and empirical calculations were called into question when they were assessed against the official accounts of the Chongqing budget and the planned repayment channels of the public rental housing financial package. To begin, the notion of an international deficit warning level applies only to national economies (and even so, the 3% limit is a relative rather than an absolute indicator). The causality of this cross-scalar relation was overlooked: Chongqing is administratively a subnational economy, and its contribution to national structural coherence is contingent on the redistributive policies – developed in part through state rescaling – of the Chinese central government (ref. Figure 3.1, Chapter 3). The fiscal revenue figures for Chongqing would therefore need to account for the Chinese central government's transfer payments to Chongqing.

Both Liu and Wu's figures did not include this financing component; in addition, Wu did not consider extra-budgetary revenue figures, a glaring omission considering the information is publicly available. Because no standard rule states that fixed asset investments need to be solely financed through the redistribution of taxpayers' monies, the issue concerning state-driven projects that involve credit is whether the projected repayment channels are feasible (more on this shortly). As such, while fixed capital investments in Chongqing have indeed soared over the past five years (ref. Figure 6.2, Chapter 6), whether they offer concrete financial returns – and in turn vindicate the state's redistributive 'investments' – is contingent on the ability of these investments to drive and benefit from the accompanying series of industrialisation projects.

The inaccuracy of Liu and Wu's interpretations is further underscored by new information on Chongqing's budgetary health and debt-repayment channels (for detailed summary see Table 7.2). According to comprehensive statistics provided by an anonymous spokesperson from the Chongqing Finance Bureau, the city's fiscal standing is actually generating surpluses because of prudent spending and redistribution from the central government (anonymous source, interview with www.cqnews.net, 24 March 2012a; author's translation). To be sure, statistics officially released by state agencies are not inherently more accurate or reliable (cf. Wang 2010; Lim 2017). What these figures offer is a benchmark from which assessments of subnational governments' budgetary discipline could be made, and it is this benchmark that contradicts Wu and Liu's assertions.

More detailed assessments of potential budget delinquency could be made after the Chongqing government took the extra step – a rare one in China's context – of providing information on the funding and repayment streams for the rental housing construction (ref. Table 7.2). The-then Chongqing mayor, Huang

Table 7.2 Emergent characteristics of Chongqing's public rental housing provision.

Proposed volume	Targeted residents and rental conditions	Financing and repayment strategy
• Original plan in announced February 2010 was to build 40 million m^2 over 10 years; but target is expected to be met as early as 2012 • Construction on 13 million m^2 and 14.25 million m^2 (approximately 219,200 units) began in 2010 and 2011, respectively • Construction on 13.2 million m^2 (approximately 237,400 units) projected for 2012	*Targeted residents:* • Homeless families or families whose current living space is less than 13 m^2 per person; and • Individuals who are employed within the Chongqing core urban area and other stipulated urban zones; and • Individuals with a monthly salary not above 2,000 yuan; or a two-person family unit with a combined income not above 3000 yuan; or a bigger-than-two person family unit with an average income not above 1500 yuan • Rental rates established at around 60% of market prices (2010 levels) • *Quota:* 50% for peasant workers, 15% for new university graduates, and 35% for existing urban residents with housing difficulties • *Key qualifier:* Individuals who meet the above conditions do *not* need to hold Chongqing *hukou* status *Rental and ownership restrictions:* • Tenants are allowed to purchase their rental units from the state at just above cost-price after residing for 5 years • Tenants opting not to buy can continue to rent so long as they meet the prevailing requirements • Tenants who purchase the units cannot re-sell them on the open property markets; units can only be re-sold to the state; no secondary market allowed for public flats.	*Operating costs:* • Preliminary fiscal injection: 30 billion yuan • Opportunity cost: estimated 30 billion yuan in fiscal revenue, if the land used (based on 2010 land use planning and land valuation) is tendered out for leasing • Bank credit: first wave of loans injected by Huaxia Bank in September 2009; new loan commitments by Bank of Communications, ICBC and China Construction Bank followed; more than 10 billion yuan injected by 2012; interest rate estimated to be around 5% • Insurance firms, Housing Provident Funds, Social Protection Funds etc.; more than 40 billion yuan raised from these sources in 2011 • On 23 April 2012, Chongqing Land Group (owned by the municipal government) obtained approval to issue 5 billion yuan in corporate bonds for affordable housing; the biggest amount approved in China at the time for affordable housing construction *Repayment channels:* • Rental income used to cover loan interest payments; rental rate (hence income) subject to (upward) adjustments, in relation to per capita GDP growth • 5% of revenue collected from annual land leasing is committed to financing public rental housing construction • An estimated 1/3 of tenants expected to purchase units; sales income will be used to cover principal repayment • Of the area already completed or currently under construction (around 40 million m^2, based on March 2012 calculations), 1/10 will be rented and/or sold for commercial use at open market prices; all sales revenue will contribute towards debt repayment • **Key strategy:** The financialisation and repayment process is predicated on a structured cost-recovery model

Sources: *Chongqing Chengbao* (9 June 2010); *Chongqing Ribao* (28 October 2010b); *Nanfang Zhoumo* (26 August 2010); *China Economy & Informatization* (18 January 2012); www.news.cn (8 March 2012); *Sina Dichan* (9 April 2012); www.cqnews.net (26 April 2012b); author's notes from participation in seminars with planners and policy consultants in Chongqing. Data compiled and translated by author.

Qifan, explains that apart from bank loans and fiscal redistribution, it was able to save on spending on the land costs which, valued at 30 billion *yuan*, would be an incurred expense if the construction was left to private producers (Huang, interview with www.news.com.cn, 8 March 2012; author's translation). Huang's explanation offers a counterpoint to Wu Jinglian's interpretation that the housing project gives 'little regard' to the origins and eventual 'returns' of fiscal expenditures. Indeed, as Huang further emphasises, the provision of public rental housing operates on the basis of market-like exchange and, by implication, does not contradict Deng Xiaoping's commitment to produce a 'socialist market economy' (cf. the integration of market logics in state governance, as discussed in Chapter 2):

> The rent to be collected is at least equivalent to the interest payments for bank loans. [...] But there are still management costs and the 80 billion *yuan* principal sum to repay, how can this be balanced? No problem, because a third of the tenants in rental housing may want to buy over the title of the flats after 3 or 5 years, when they do so, at a capital cost of around 3000 *yuan* per m^2, it is possible to repay 40 to 50 billion *yuan* with sales of more than 10 million m^2. In addition, 4 million m^2 of the 40 million m^2 will be sold for commercial use…[it] will generate 40 billion yuan, so this debt can be balanced out. (Huang, interview with www.news.com.cn, 8 March 2012; author's translation)

Of particular significance in the repayment plan is the function of land as *at once* a saleable commodity and a public good. This was possible because weaker market reforms relative to the coastal city-regions invited sustained economic intervention by the Chongqing government (as discussed in Chapter 6). State functions were consequently reconsolidated to become more market-like while keeping intact its ability to determine the exchange value of land *as and when is required*. In the instance of housing provision for the lower income populace, land valued at 30 billion *yuan* was stripped of its price (i.e. exchange value = 0 *yuan*). This redistributive process illustrates how the CPC remains 'locked in' to the path of public landownership instituted by Mao in the 1950s. Its ability to attach a monetary value to land exists because the CPC never relinquished its control of land. This phenomenon is more pronounced in regions like Chongqing that were less exposed to the industrialising reforms of the 1980s and 1990s.

As Liu Guoguang, the former Deputy Head of the Chinese Academy of Social Sciences, puts it, the Chongqing government was well-placed to take on this challenge because it retained a crucial precondition of effective redistribution – public ownership (Liu, interview with *Chongqing Ribao*, 5 August 2011; author's translation). Contrary to Mao Yushi and Wu Jinglian, Liu's logic arguably underpinned the Chongqing reforms. Through strong public ownership, the Chongqing government was able to prioritise use value over exchange value in cases it deems to be more socially effective. It was therefore able to rely exclusively on SOEs in the construction of the public rental flats. Whether this prioritisation is less economically 'efficient' is a non-issue because, to follow Mao

Yushi's aforementioned acknowledgement, market mechanisms inherently generate and are unable to overcome social disparities (cf. *Shenzhen Special Zone Daily*, 4 June 2012).

The interesting characteristic of the Chongqing reforms is the incorporation of market logics in the state's calculations. This is clearly a legacy of the reconsolidation of Chongqing SOEs in the early 2000s (ref. Chapter 6). Plans were implemented to enable the government to demonstrate its ability to recover the costs 'invested' in the social facilities, which in turn allowed it to avoid charges of financially irresponsibility. That Huang acknowledged the importance of bank credit as a financing channel is unsurprising; as Qiu Feng has emphasised, debt-financing commonly underpin subnational developmental agendas across China (cf. framework presented in Chapter 3). Of theoretical significance is Huang's detailed explanation of how the 'debt can be balanced out'. Specifically, the entwinement of the 'balancing' process with market mechanisms is telling. As Table 7.2 shows, the rental rates are not fully subsidised by the state; rather the costs can only be recovered because part of the available real estate would be rented or sold at market-determined prices. The redistributive process of converting land into a public good is thereby contingent on the state's ability to perpetuate its characteristic as a commodity. If anything, this illustrates a tension that is intrinsic to the CPC's official pursuit of a 'socialist market economy'.

Through the analysis of the funding rationale, the state-driven attempt to alter the spatial logic of socioeconomic regulation in Chongqing cannot be framed dichotomously as the 'state vs. the market' or 'politics vs. economics'. Viewed within the context of the analytical framework presented in Chapter 3, it is about the *interaction* between the state apparatus, transnational circulatory capital and a subnational developmental agenda; it is about a longstanding regulatory legacy whereby politics determined socioeconomic 'development' through spatial reconfiguration (as first introduced in Chapter 1, second section). As such, the seemingly high-cost investments to effect urban–rural integration in Chongqing cannot be taken as antithetical to deepening marketisation – or, more precisely, to deepening state-led marketisation. The very raison d'être of urban–rural integration is to facilitate the socio-spatial absorption of peasant migrant workers that would in turn support the incoming industries (a substantial number of which are state-invested). Because market mechanisms have become the preferred tools to regulate social life, what will concretely define 'development' in China will inevitably involve market exchange. This aspect is arguably the primary path-changing characteristic of policy experimentation in and state rescaling through Chongqing. In the context of this book, the crucial questions generated by this characteristic are whether (i) the integrated developmental approach in Chongqing contradicts the rationale of state rescaling and policy experimentation in Guangdong; and (ii) whether the seemingly variegated expressions of state rescaling across China are an intended effect.

Conclusion

> One test for whether reform is genuine is if people will emerge screaming about the
> pain they feel as a result. If no one feels any pain, that is not reform. – Hu Shuli,
> Editor-in-Chief of Caixin Media. (Interview with *Asahi Shimbun*, 18 April 2013)

Much has been written on the 'Chongqing model' or 'Chongqing experience' of development since the Bo Xilai government launched a broad series of socioeconomic reforms in late 2007. Yet there remains an empirical gap when it comes to illustrating the causes and implications of these reforms. The research presented in this two-chapter segment is an attempt to add clarity to this growing literature. Through establishing the connections between the post-2007 experimental reforms, which are ongoing, and their conditions of possibility, some of which could be traced back to the Mao-era, these two chapters have shown that the 'scaling up' of the reforms was never an expression of linear historical evolution. In contrast to the export- and market-oriented developmental approaches in Guangdong, the site of Deng's first-wave experimental reforms, the Chongqing government had been strengthening its interventionist hand since the early 2000s (ref. Chapter 6, second section). Responding to and building on this strong state capacity, the former Chongqing Party Secretary Wang Yang began to work closely with the central government to launch experimental policies to reform the 'dual structure'. These developments did not generate accusations that Chongqing represented a (re)turn to the Cultural Revolution, however. Reforms in Chongqing took on a whole new level of prominence following the global financial crisis and the subsequent push to designate Liangjiang New Area as 'nationally strategic'.

This chapter has analysed how two interrelated socioeconomic reforms to actualise urban–rural integration were justified as integral to the industrialisation drive in Chongqing. These reforms are, namely, the large-scale public rental housing provision and the equally large-scale conversion of peasant migrants' *hukou*. On their own, these reforms would arguably not have attracted much attention: public rental housing provision was not a novelty then, and there was talk of overturning the urban–rural 'dual structure' for a long time (ref. Chan and Buckingham 2008). It was the justification of these reforms as fundamental to industrialisation that generated the strong debates. As the US subprime mortgage crisis expanded into a global financial crisis in 2008, the Chongqing government went on a discursive offensive: it began to argue that it is possible to perpetuate and proliferate industrialisation (spearheaded by the Liangjiang New Area) without exacerbating the socio-spatial inequality that already exists at the national scale. This integrated approach underscores a fundamental difference between the Chongqing reforms and earlier state spatial strategies of earlier regimes – it very clearly showed these earlier approaches to have decidedly overlooked issues pertaining to 'fairness', 'justice' and 'citizen rights' (ref. former Chongqing mayor Huang Qifan's comment earlier in this chapter).

It is thereby more accurate to construe the reforms in Chongqing as at once an outcome of and a response to the economistic developmental approach that characterised the post-Mao era. It is almost as if the central government pulled open a space to foreground and resolve the constraints of this approach. When the Special Economic Zones were instituted in 1980, the necessity to ease socio-spatial inequality did not exist. The conditions that drove SEZ-formation were primarily generated by Mao Zedong's decision to engage the global system of capitalism (set in motion by the Nixon visit to Beijing in 1972). Ironically, however, Deng kept Mao's 'dual structure' institution and established it as the *spine* of subsequent spatial strategies. In turn, the reforms in Chongqing became a necessity because the 'ladder step' strategy instituted by Deng and his successors generated colossal inter-provincial disparities and exposed the social limits of the 'dual structure' under conditions of nationwide population mobility. This relational vantage point brings into sharper focus why the 'Chongqing experience' generated nationwide debate: the reforms directly question whether the post-1949 approach to integrate Chinese economy and society on the basis of institutionalised uneven development was fundamentally flawed from its very beginning.

Interestingly, as discussed in Chapter 6, the strategies to experiment with new logics of socioeconomic regulation in the first two decades of the post-Mao era left open room for governments in less developed regions like Chongqing to reconsolidate their economic strength so as to drive regional development. Of particular significance is the incorporation of market mechanisms to reform the urban–rural 'dual structure'. While the Chongqing government involved only SOEs in the construction of public rental flats, these SOEs (e.g. Chongqing Land Group) were able to raise funds through credit and equity markets located in different locations. By unilaterally establishing the exchange-value of rental apartments, the Chongqing government could undercut what private producers are otherwise expected to charge. Property speculation was further precluded through the prohibition of a secondary resale market for rental flats (cf. Table 7.2). Yet the cost-recovery model indicates the social reforms do not constitute a direct subsidy; on the contrary, the act was justified on the basis that it would allay investors' worries about finding accommodation for workers. This further accentuates the tight state-market entwinement in Chongqing: state ownership of a sizeable segment of the housing market (i.e. that consisting of low-income earners) was paradoxically financed by (nominally-) private banks and private economic investors, in the name of facilitating expanded capital accumulation. Unprecedented in China, this integrated developmental approach arguably placed other provincial and municipal governments under pressure: it underscored what the CPC did not do in earlier rounds of industrialisation, that is offer social benefits to the growing number of migrant workers in industrialising city-regions.

The decision to launch the '314' agenda in a geographically-targeted location (i.e. Chongqing) indicates how reforming seemingly intractable institutions of the Mao era cannot be easily implemented nationwide. As the varied responses to the reformist policies indicate, this agenda simultaneously pushes against and is hemmed in by the broader state spatiality in which it is located. The 'hemming in' process is especially pronounced when compared to reforms occurring simultaneously in Guangdong (ref. Chapters 4 and 5): the very same person – Wang Yang – who sought to reform the 'dual structure' used precisely this structure to actualise the 'double relocation' of industries and labour. This ironic phenomenon reveals another characteristic of the Chinese socioeconomic developmental approach: institutionalised spatial unevenness has become of constitutive value for the regulation of Chinese economy and society, so much so that generating a new regulatory path is fraught with difficulties.

While this path-dependency generated strong opposition to the quest for urban–rural integration in Chongqing, the very emergence of opposition strongly suggests vested interest groups have come to dominate economic development in China. True reforms, says the *Caixin* editor Hu Shili, trigger pain and screams. The 'screams' in response to the Chongqing reforms illustrate how select groups have benefited from national socioeconomic policies instituted as far back as 1958. The reason why intense opposition emerged almost right from the beginning of the reforms in Chongqing is thereby symptomatic of a fear that the CPC was genuinely prepared to remove the 'primary obstacle' to urban–rural integration. Yet the emergence of this fear reflects the spatial potential internal to the pursuit of spatial egalitarianism in Chongqing. Its existence suggests that fresh developmental undercurrents are forming; that the future of Chinese economy and society could have a new institutional foundation.

In this respect, the socioeconomic reforms in Chongqing presented in this two-chapter segment contradict conventional portrayals of 'Maoist' resurgence in the municipality. Contrary to Mao's favoured dialectical approach of launching an 'affirmative' event through the 'negation' of another separate event, the Chongqing government had effectively shown that, with effective redistribution, 'affirmation' (economic growth through large-scale industrialisation) need not be based on 'negation' (the 'dual structure' of uneven development). If anything, the socioeconomic reforms instituted in Chongqing are in part a form of anti-Maoism cloaked in the spirited melodies of the municipal government's favourite 'Red Songs'. What makes it distinct as a 'model' – and which also contrasts strongly with the 'double relocation' and financial policy experiments in the Greater Pearl River Delta – is its objective to achieve simultaneous economic growth and surplus redistribution. It is for this reason that the Chongqing experiments stand distinct from the rest of the country.

This increasing importance of spatial justice on the marketisation agenda is similar, interestingly, to the CPC's developmental approach prior to the 1958 'socialist high tide'. Between 1950 and 1952, the CPC centralised and facilitated the egalitarian redistribution of a 'fictitious commodity' – land – for poor peasants across the newly established nation-state. Through the Land Reform Act known colloquially to have 'altered the heavens and changed the earth' (*gaitian huandi* 改天换地),

the entire populace experienced common affluence for the first time in the history of 'new China'. Integrating the national economy on this egalitarian basis, the land reforms arguably constituted the first socially-progressive event launched by the CPC. As the contemporary Xi Jinping leadership contemplates unleashing a new wave of economic growth through nationwide urbanisation, the experimental quest for spatial egalitarianism in Chongqing suggests that, if the proposed urbanisation process unfolds on the basis of social equity, Mao's 1956 vision of 'the newest and most beautiful painting' could yet (finally) be in the making.

Endnotes

1 While the post-2007 per capita gross income in Chongqing have risen significantly, the disposable income of rural residents remained relatively low at ~5278 *yuan* per capita in 2010. Per capita disposable income for residents with urban *hukou* was ~17,532 *yuan* (*Chongqing Shangbao*, 28 January 2011. Comparatively, per capita disposable income for Shanghai's urban residents was 31,838 yuan in 2010, while rural residents earned 13,746 *yuan* per capita (www.eastday.com, 25 January 2011).

2 Most real estate developers, while incorporated, are either state-owned or joint ventures involving state partners (of different levels of the administrative hierarchy). For this reason, they are 'nominally' private, but are in reality entities related to the Chinese state apparatus. Because this state-economy overlap produces different economic groups, all with separate profit-seeking interests, it is often said that reforms in China has to negotiate a labyrinth of 'vested interest groups' (*liyi jituan* 利益集团).

3 Comparatively, National Bureau of Statistics of China data puts public rental housing at an estimated 7% of the national urban housing stock in 2007 (see Man 2011). Assessed vis-à-vis growing urbanisation and housing price inflation, it is clear that following nationwide housing reforms, there has been a marked shift towards the private provision of residential housing.

4 After Deng launched China's economic reforms, housing officially became a commodity in China in 1988. Since 1998, state-owned work units were prohibited from directly providing housing for workers, which expanded private housing markets. (For more elaborate discussion, see Zhou and Logan 2002; Huang 2004.) Together with the effect of surplus capital looking for new investment outlets, prices for privately-supplied housing have experienced an upward surge since the mid-2000s. This consequently produced what the *New York Times* (14 April 2011) called 'China's scary housing bubble'.

5 At the onset of reforms in 2010, prices of market-supplied housing in Chongqing was around 6000 *yuan* (~US$950) per square metre, almost six times lower than the 35,000 *yuan* average in Beijing. Because the Chongqing government still retained strong control of land resources, it was in a driving position to determine housing supply by intervening in the rental housing market. Whether it could still do so after leasing out its land supplies to private developers remains an open empirical question.

6 This figure does not apply to shareholdings, which could involve SOEs, collectives and joint ventures. If these figures are included, the overall proportion would reach around 60% (Chongqing Bureau of Statistics 2013).

Chapter Eight
Concluding Reflections

Introduction

The relationship between the 'feeling for stones' regulatory approach and foundational institutional change in post-1949 China has been under-explored in the interdisciplinary field of political economy. How geography – or, more precisely, geographically-differentiated regulation – plays a role in this relationship is a pronounced research gap. Amalgamating and building on conceptualisations of state rescaling, policy experimentation, and path-dependency, this book has presented an explanation and evaluation of an increasingly distinct geographical trend in the CPC's new attempt to 'feel for stones' – the designation of 'nationally strategic new areas' to facilitate policy experimentation. The analytical framework was developed on the assumption that the emergent territories of Chinese state rescaling constitute an empirical lens through which to evaluate the stability of regulatory foundations in post-Mao China. Three dynamic parts constituted the narrative: first, the study developed a thorough geographical–historical reassessment of the *rolling* spatial regulatory approaches instituted since the formation of a Chinese nation-state in 1949. The second part then established the framework for empirical research that explicitly considers the role of transnational circulatory capital and subnational developmental agendas on national-level regulation in post-Mao China. This framework sets up the third part, namely

On Shifting Foundations: State Rescaling, Policy Experimentation and Economic Restructuring in Post-1949 China,
First Edition. Kean Fan Lim.
© 2019 Royal Geographical Society (with the Institute of British Geographers).
Published 2019 by John Wiley & Sons Ltd.

an empirical examination of the rationale and effects of designating 'nationally strategic new areas' in the Pearl River Delta (Chapters 4 and 5) and Chongqing (Chapters 6 and 7).

As the analysis illustrates, the retention of national structural coherence in China is inextricably intertwined with the reconfiguration of territories to facilitate differentiated policy experimentation. Emphasis was given to the *tensions* between path-dependency and path-generation in the post-Mao era. Two major waves of changes transformed socioeconomic and political life from what it was like in the pre-1949 'old society'. The first wave unfolded across the national scale and consisted of land redistribution, enforced collectivisation of production and demographic control through the *hukou* system. The second wave was rolled out at 'meso' levels through reconfiguring production and mobility regulations in the rural hinterland and selective engagement with the global economy through targeted city-regions. The first wave was oriented towards and modelled after the Soviet political economy, the second was oriented towards (but not necessarily modelled on) the (neoliberalising) global economy. Because the outcomes of the first wave could not align with the global neoliberal ideology, the most recent round of policy experimentation constitutes a simultaneous attempt both to resolve the contradictions of the second wave reforms and to repurpose the logics of socioeconomic regulation from the Mao-era. It is therefore more appropriate to think of these rounds of regulation, each with their own geographies, in terms of a complex palimpsest rather than a straightforward process of succession (Chapter 2).

Specifically, the comparison of the characteristic spatial form of the Mao-era and that of the post-Mao 'transition' problematises binary conceptualisations of politico-economic evolution in post-1949 China. The two case studies in this book have reinforced the argument that the centralisation–decentralisation and spatial egalitarianism–uneven development binaries do not explain (i) entrenched uneven development in the Mao era (albeit at much lower income and output levels); (ii) the reconfiguration of central state power since the mid-1990s; and (iii) qualitative changes in central–local relations. Path generating initiatives such as the institution of city-regional policy experimentation may have been driven by state rescaling, but they continue to co-exist with seemingly intractable regulatory pathways such as the urban–rural 'dual structure' and the coastal concentration of industries. The outcome is therefore a nuanced picture of national structural coherence as previously introduced in Chapters 1 and 3: new rounds of regulation may have produced new changes, but traces of earlier forms remain.

Placed in relation to the research agenda presented in Chapter 3, the book arrives at five interrelated conclusions. The first two concluding points address questions (a) and (b), and the subsequent three points address questions (c) and (d):

• *Contrary to the Mao era, there is much greater spontaneity and spatial selectivity in the ways initiatives of national significance are proposed, evaluated and ultimately implemented.* While various forms of decentralised governance characterised both eras, the regulatory rationale in the Mao-era was to keep each province (and the respective administrative scales within each province) separate and directly under the control of the central government (ref. Chapter 3). While this regulatory logic still impacts inter-provincial relations in present-day China, the provincial governments now have greater agency in putting forth developmental agendas in the name of the 'national interest'. This in turn engenders strong competition between provinces for the coveted power to 'move first, experiment first'. Crucially, this competition increases at once the power of the provincial and central governments. For the provincial government, each administrative locality had to convince the Party Secretary and Governor of his/her jurisdiction's ability to be the next 'nationally strategic new area'; this adds further dynamism between the relations between the provincial and the county/municipal governments. The same logic applies to central-provincial relations. Viewed in relation to Figure 3.1, the competition generates the required subnational responsiveness to sustain national-level structural coherence. That decentralised governance has taken on this effect of enhancing the legitimacy and power of the central government against a globally interconnected context is an important characteristic that the 'centralisation–decentralisation' debate failed to illustrate (ref. Chapter 2). And this leads to the next concluding point.

- *State rescaling and geographically-targeted policy experimentation across China have become integral to the preservation of domestic socioeconomic stability vis-à-vis an increasingly volatile global context.* Read in relation to Chapters 1 and 3, the 'double relocation program' in the Pearl River Delta (PRD) is a product of the export-oriented developmental strategy of the early 1980s. With the industrial structure strongly entwined with global production networks, the attempt at industrial upgrading had to ensure any disruption to this entwinement would not strongly undermine the PRD's articulation in the global economy. The goal of designating Hengqin and Qianhai as national centres of financial reforms effectively allows the CPC to reproduce, if not deepen, global economic integration (Chapter 5). In Chongqing, the repurposing of state participation in the economy vis-à-vis (the lack of) private capitalist forces allowed the CPC to create new manufacturing bases that in turn ensured the Chinese macro-economy was not strongly affected by industrial upgrading along the coastal seaboard (Chapter 6). With local governments across the country submitting proposals that potentially address national-level regulatory issues, the CPC could address different problems exposed by the global financial crisis, namely its over-exposure to the US dollar and the growing cost pressures on manufacturing along the coastal city-regions. The open question is whether 'nationally strategic' experimental policies need eventually to be extended to the national scale, as Heilmann and Perry (2011) suggest, or whether the geographical specificity of policy experimentation is *precisely* what is required for adaptive governance.

- *The main difference between the policy experimentation in the Pearl River Delta (particularly Hengqin and Qianhai New Areas) and Chongqing (Liangjiang New Area) is the generation of geographically-distinct regulatory paths in the name of the 'national interest'.* The ongoing policy experimentation in Hengqin and Qianhai are intended to enhance the internationalisation of the RMB, which in itself was initiated in response to the Chinese state's over-exposure to the dollar (as reflected in its massive and still growing dollar reserves). The experiments in Chongqing illustrated two other issues pertaining to national cohesion: the inherently discriminatory *hukou* institution and the coastal-oriented industrialisation approach. These different 'models' or approaches are therefore more accurately *extensions* of regulatory issues confronted by the Chinese central government. As Figure 3.1 illustrates, however, the state rescaling process is situated within and has to negotiate with inherited developmental pathways. It is for this reason that the apparent competition between these models/approaches could not have been fully anticipated. Rather, these variegated regulatory geographies are *simultaneous* outcomes of inherited policies and targeted responses to the problems generated by these strategies.

- *Policy experimentation in the contemporary 'new areas' is filled with uncertainties and inconsistencies, which underscores the lack of a coherent 'national strategy'.* In the case of Hengqin and Qianhai New Areas, the financial reforms were launched vis-à-vis actually-existing channels of unregulated cross-border capital flows (Chapter 5). Without reducing these channels, offshore RMB could only flow to locations designated by the CPC or, as is already happening now, move informally into and out of the Chinese economy through quasi-clandestine means. How the reforms in Hengqin and Qianhai contribute to the 'liberalisation' of the entire Chinese financial system thus remains very unclear. In Liangjiang New Area (and Chongqing more broadly), the prominent roles of fiscal redistribution and local state financing generated strong resistance by prominent actors. While not explicitly stated, this resistance might have contributed to the lack of enthusiasm for similar reforms in other provinces, which in turn highlights how developmental approaches across China are entwined with the regulatory path established by the 1958 urban–rural 'dual structure'. The defining characteristic of this path is the institutionalisation of the population into 'agricultural' and 'non-agricultural' classes and, on this basis, the facilitation of urban-based industrial growth through the extraction of rural resources (the most important of which is rural labour power). While the current Xi Jinping administration remains committed to nationwide equitable urbanisation, the repeated delay of the promised blueprint strongly suggests the 'socialistic' approach in Chongqing is not applicable across the wider Chinese political economy.

- *Designed to fit the pre-existing socioeconomic conditions of Guangdong and Chongqing, the experimental reforms generated contradictions that highlight the difficulties of extending them 'as is' to other locations with different socioeconomic developmental pathways*. If both the Guangdong and Chongqing approach to socioeconomic restructuring were to be extended nationwide, the narrow focus on industrial restructuring (and, by extension, the perpetuation of GDP growth) in Guangdong would undermine the socialistic approach in Chongqing. In Guangdong, as Chapter 4 shows, the Wang Yang administration chose to drastically amend the industrial foundation of the Pearl River Delta while it set up the designation of Hengqin and Qianhai into 'nationally strategic' areas. The aim of state rescaling was predominantly economistic: it was to ensure that Guangdong retain its leading economic position in China. In Chongqing, however, the Bo Xilai administration had to build on the equitable urbanisation pathway established by the former Wang Yang administration while it sought to attract transnational circulatory capital. The outcome is an integrated developmental approach that is called 'socialistic'. State rescaling and policy experimentation in China thereby generate divergent developmental pathways, such that no single model seems likely to predominate at the national scale.

The following sections of this chapter will provide critical reflections on these five conclusions. The next section situates the main conceptual arguments of the book in relation to the key empirical contributions summarised above and assesses the extent to which institutional foundations have truly shifted in the Chinese political economy. The final section highlights and assesses the new questions that emerged in the course of this study, with a view to developing them in future research projects.

Overview of Arguments and Conceptual Contributions

As mentioned previously in Chapters 1 and 3, the primary conceptual question for understanding state rescaling in China is whether there were similarities in the triggers of rescaling relative to those of the Fordist heartlands (and hence the growing focus on devoting developmental resources to city-regions). Through the analysis in Chapter 2, the book introduced the conditions and tensions that generated qualitatively new forms of decentralised governance in the 1980s. This geographical–historical re-appraisal indicates how the nationally-oriented and geopolitically-insular regulatory logics of the Mao era generated crisis tendencies that impelled the new Deng Xiaoping administration to launch a geographically-targeted policy experimentation through the 1980s. The primary contrast between the Chinese context and the western European cases is that crises in China were not solely a reaction of the 'first wave of urban locational policy'; in China, the growing prominence of locally-driven developmental initiatives since the early 1980s was in part a reaction to the excessive emphasis on communal production in the vast rural hinterland and in part the emphasis on place-based policy experimentation with marketisation. Indeed, the experimentation with the Household Responsibility System in the rural hinterland marked the fulfilment of a key restructuring goal devised by senior Chinese policymakers of the 1960s (e.g. Liu Shaoqi, Li Fuchun and Deng Zihui), that is, to institute a regime of production across the rural communes that allows individual households some autonomy in production.

Despite these reformist policies, the very logics of socioeconomic regulation that generated the 'miraculous' economic growth of post-Mao China contain residues of Mao-era logics. As introduced in Chapter 1, experimental approaches to state formation in contemporary China are not unique to the post-Mao era. Prior to securing power in 1949, the integration of disparate economic geographies (e.g. Manchuria, Tibet, Inner Mongolia and smaller fiefdoms) into a national entity was predicated on the localised mobilisation of peasants to overthrow exploitative landlords. The CPC's current ability to renegotiate central-local regulatory relations is thus a Mao-era legacy. This began with a politically orchestrated strategy to expand support for the CPC takeover of power through the quest for egalitarian land redistribution. Once power was secured, the Mao administration fulfilled its promise through the Land Reform Act. But this was short-lived: on the premise of 'national economic construction', the Mao administration launched and jettisoned several programs – land and other means of production were nationalised in 1958; the transfer of regulatory power to six administrative regions in 1949 was abandoned in 1954; and the Great Leap Forward industrialisation program, launched in 1958, was stopped in 1960. What ensued was the Third Front construction campaign, followed by the Cultural Revolution; means of production were moved around, only to exacerbate uneven economic–geographical development (Chapters 1 and 2).

While the individual campaigns generally did not last long, the three major policies that enabled Mao's state-building process – land nationalisation, the People's Communes and the *hukou* institution – endured at least two decades. Of the three, only the People's Communes were disbanded in 1984 as greater freedom for individual production was granted; despite multiple land reforms and, in the context of this book, the experimental policies to overhaul the *hukou* institution in Chongqing, the CPC *continues* to launch reforms on the basis of a nationalised land system and the urban–rural 'dual structure'. Establishing the logics and impacts of these institutions was important for two reasons. First, they offered an insight into what has been a less-discussed precondition of post-1978 reforms in China: *the consolidation of central political power through decentralised governance*. Mao-era spatial logics of regulation – expressed through the People's Communes, urban industrial work units and the 'cellular' relationship between these units – allowed the Chinese central government to fortify its political control over the newly-formed state spatiality. And it was this control that in turn allowed successive governments, led by Deng Xiaoping and Jiang Zemin, to launch more spatially-targeted strategies for socioeconomic regulation (ref. Chapters 1 and 3).

Second, extending the respective legacies of these three policies to the present enabled a clearer theorisation of the rationale of contemporary state rescaling (expressed in/through the demarcation of 'nationally strategic new areas' for policy experimentation). At one level, the Mao-era regulatory logics that defined CPC rule have become increasingly blurred – if not destabilised – by intensifying flows of capital, labour and commodities within Chinese state space. This instability was particularly pronounced in the immediate period following the global financial crisis in 2008. For Hu Jintao, the former Chinese president who intensified the extent of 'nationally strategic' rescaling after 2008, this re-assessment became a pressing policy task. The challenge to stabilise the Chinese macroeconomy was made tougher by the need to address the combined socioeconomic effects of the two preceding waves of spatial changes and increasingly volatile global economic conditions. The outcome was a further fragmentation of Chinese state spatiality into institutionally-distinct regulatory spaces known as 'nationally strategic new areas', with each area responding 'strategically' to 'national' challenges by integrating with the global economy in their own ways. After Hu passed the regulatory baton to Xi Jinping in 2013, this approach has expanded in scale and scope (ref. Chapter 1).

This said, the Chinese central government never fully relinquished its control of the means and output of production after the Mao era (Chapters 2 and 3). That the CPC requires *sustained* control over production arguably explains why tensions emerged from state rescaling in the Pearl River Delta and Chongqing. On the one hand, established policies generated gains for interest groups. Resistance to change unsurprisingly ensued vis-à-vis the instituted relocation of unwanted industries out of the Pearl River Delta (Chapter 4) and the implementation of equitable urbanisation in tandem with the launch of Liangjiang New Area in Chongqing (Chapter 7). These tensions illustrate a key characteristic of state rescaling in China: policymakers at lower administrative levels are given flexibility to propose and implement experimental policies of potential significance to the 'national interest' – a significant change since the 'cellular' mode of control during the Mao-era – through expanding the pool of stakeholders; yet, in pushing through these new forms of adaptive governance, these policymakers have to confront their own party's reluctance to alter specific policies deemed fundamental to the preservation of CPC rule. This development in turn underscores the malleability of institutional foundations across China: the contemporary globalising context has engendered a whole new array of regulatory challenges that are different from those confronted by the Mao administration as it seized control from the KMT in October 1949.

What has changed, specifically, is the experimental approach to confront these challenges. Contrary to the form of adaptive governance defined by Heilmann and Perry (2011), that is, the expansion of small-scale experiments to the national scale if the experiments are deemed to be beneficial across the country, the emergence of new regulatory forms in the 'nationally strategic new areas' reflects a dialectical process in which decentralisation becomes a *function* of centralisation, or decentralisation *as* centralisation. Integral to this process is the direct *national-level* coordination of place-specific policy experimentation rather than the extension of experimental policies nationwide. State rescaling therefore allows the Chinese central government flexibility to cushion and respond to new developmental challenges/pressures by making use of *geographically-specific* conditions (e.g. the relatively high degree of economic integration between the Pearl River Delta and the two open economies of Macau and Hong Kong, the huge rural population within the expanded Chongqing municipality, etc.). To be sure, the scalar shifts in political power in and through the 'new areas' illustrate the contingent and capricious nature of the conditions that made possible flexible and sustained central control. As the two case studies in this book demonstrate, the establishment of new regulatory scales produced regionally differentiated and seemingly heterogeneous developmental pathways that rendered unclear the plausibility of extending experimental policies nationwide. It might be more plausible, indeed, for experimental learning to be 'lateral', that is, it gets extended in a local-to-local manner.

A particular problem emerges when the competitive impulses to copy and replicate experiments in one place generate new crisis tendencies within the Chinese political economy rather than successfully reconstitute central state power. In itself, the state rescaling process exemplifies and is an outcome of two recurring contradictions. First, the production of Hengqin, Qianhai, and Liangjiang New Areas into new state spaces of socioeconomic regulation underscores how national economic–geographical control is secured through destabilising existing administrative boundaries. In other words, coherent national-scale governance is contingent on the generation of new subnational borders. That this re-bordering is not novel strongly suggests it has become a necessary precondition of China's engagement with the global economy (cf. Ong 2004, 2006). Second, the case studies showed how the production of new 'nationally strategic' regulatory spaces does not imply a total break from institutions inherited from as far back as the early years of CPC rule. Relative to developments in the Fordist heartlands, the emergent shift to city-regionalism is more explicitly concerned with sustaining national-level structural coherence. Indeed, the state rescaling outcome appears to *repurpose* these institutions – specifically, centralised financial and demographic control – in ways that would allow the economic–geographical structure of China to remain competitively articulated to the global economy. How – or whether – place-specific policy experimentation could pre-empt or contain crisis tendencies would thus be an important focal point in future research agendas on socioeconomic regulation in China.

A further conceptual point emerging from the preceding analysis is the difficulty – if not impossibility – of considering whether the Chinese political economy could be successfully 'reformed' vis-à-vis a universal template (e.g. a prototypical 'neoliberal state' or a Marxian 'socialistic' end-state). Rather, it would be more fruitful to adopt a grounded approach that explores how the CPC reconfigures space in order to preserve its power within the context of a crisis-prone global economy. As this book has shown, the spatially-targeted reconfiguration of national regulatory capacities allows it to remain sensitive to global economic instabilities. Through integrating Hengqin and Qianhai as functional extensions of the financial industries in Hong Kong and Macau, central regulatory authorities appear to be experimenting with new spatial conduits to undergird the RMB internationalisation program. This internationalisation intention became clear after the global financial crisis deepened in 2008, with then Premier Wen Jiabao pledging to reduce the usage of an unstable US dollar and encourage wider cross-border trade settlements in the RMB. The Guangdong government grasped the opportunity to support the internationalisation drive, as discussed in Chapter 5, leading in turn to the 'upgrading' of Hengqin, Qianhai, and Nansha New Areas into 'three strategic chess pieces'.

Yet the important role of the central government in co-determining, both politically and economically, the state rescaling agenda in the PRD and Chongqing reinforces the argument introduced in Chapters 1 and 3, namely that economic liberalisation in China is ultimately predicated on the *re-negotiation* rather than a linear, one track movement of regulatory relations within the administrative hierarchy. This process also includes the extra-territorial governments of Hong Kong and Macau, as the Hengqin and Qianhai case study shows. The Chinese central government's preference for geographically-targeted policy experimentation rather than total regulatory devolution (i.e. the permission of total liberalisation nationwide) in turn suggests there is no intention on the part of the CPC to 'liberalise' the Chinese financial system in toto. Understanding domestic financial reforms is especially pertinent in the context of RMB internationalisation. As the policy experimentation in Hengqin and Qianhai are still ongoing, the impacts of new policies will continue to invite future research.

The nationwide debates generated by the socioeconomic reforms in Chongqing – and, more crucially, the inability to expand these reforms nationwide – illustrate the difficulties of reforming the Mao-era *hukou* institution and the corresponding urban–rural 'dual structure' (Chapter 7). From another angle, it could be argued that this difficulty emerged because many interest groups – SOEs, TNCs, village collectives located around coastal city-regions, etc. – have accumulated capital on the back of this institution in the post-Mao era (cf. discussion of 'price scissors' and collectives' complaints in Chapter 2). As Chapter 4 shows, the Guangdong government's ability to launch swiftly the 'double relocation' program was due in part to the large concentration of migrant workers without local *hukou*. Because these migrants remain legally obliged to return to their places of *hukou* registration (*hukou suozaidi* 户口所在地), they were inherently in a subservient position vis-à-vis the Guangdong government. Dismantling this inherently discriminatory 'dual structure' nationwide would thus entail challenging interest groups that benefit directly from this structure in the process of capital accumulation and socioeconomic regulation.

Interestingly, as the proactive engagement with transnational circulatory capital in Liangjiang New Area demonstrates, an integrated mode of development is achievable in reality without compromising GDP growth (Chapter 6). After all, the CPC is supposed to stand apart from parochial economic interests and implement policies for the social welfare of the masses. The problem, however, is that many CPC members are simultaneously embedded in these interest groups; political and business interests overlap in a manner unique to the Chinese macro regulatory system (Dickson 2008; Hsueh 2011). Expanding the Chongqing reforms across the country would thereby mean unsettling the very 'dual structure' that enabled the CPC to deliver the economic 'miracle' of the past 30 years. The Hu administration tried, but could not resolve this precondition of economic growth in post-Mao China (Chapter 7); it could not overcome the fact that growth took place on the back of instituted uneven development. Whether Xi Jinping could do so with his proposed blueprint to intensify nationwide urbanisation remains an open empirical question.

What this book's research on state rescaling and policy experimentation has shown, then, is that the elevation of city-regions to 'nationally strategic' status through resource mobilisation at *both* subnational and national levels has become a primary regulatory process within the Chinese party-state apparatus (ref. Figure 3.1, Chapter 3). This is an aspect that the centralisation–decentralisation debate did not adequately conceptualise and on which further research needs to be conducted. Political actors positioned at lower levels of the administrative hierarchy need the power of the central government to re-draw boundaries and raise funds (through the state-owned financial system), while the central government seeks to benefit from favourable economic–geographical contexts to implement national-scale institutional reforms. The interaction between these actors thus allows new institutions to be formed (in the zones newly-designated as 'nationally strategic') without generating shocks to the national regulatory system. This said, it is yet unclear how the 'success' of rescaling is to be defined or measured. Whether policy experimentation in specific socioeconomic contexts could – or should – be successfully extended across the country is a theoretically-significant empirical question: the answer(s) will indicate whether adaptive governance, defined as an expansion process where small experiments rise and spread outwards to the entire country, enables coherent regulation of the national political economy.

Taken together, the findings from the empirical case studies became the basis for a macro geographical–historical re-assessment of the spatial logics of socioeconomic regulation during the Mao era. Specifically they enabled new questions to be asked of Mao-era regulatory logics and paved the way for a reconstructed historical narrative that is engaged with contemporary debates of state rescaling in China. As introduced in Chapter 1, 'China' was a restless space before the

CPC took power in 1949: territorial contestations between colonialists (Japanese and Soviet), local warlords and competing political parties (the dominant two being the Kuomintang and the CPC) were de jure [legally recognised]. One of Mao Zedong's distinct contributions to contemporary China was his ability to integrate – through his fluid notion of egalitarianism – what were otherwise disparate socio-economies. The reconstructed historical narrative in Chapter 2 thus established the platform for conceptualising why Mao-era spatial strategies triggered state rescaling and policy experimentation in the post-Mao era. As these chapters contend, some institutional continuity persists vis-à-vis change attempts precisely because of the need to ensure the central state apparatus retains the power to determine state spatial stability and security in the final instance.

The geographical–historical sensitivity to the connections between the two eras also addressed the difficulty of periodising politico-economic evolution in post-1949 China (ref. Chapters 2 and 3). To be sure, it is an epistemological requirement to develop conceptual categories to aid understanding, and temporal categories are often adopted to frame discussions. It would be helpful, however, to eschew categorising on the basis of linear historical evolution. As the experimental policies in Hengqin, Qianhai and Liangjiang New Areas reveal, overturning regulatory logics inherited from the Mao era across the national scale has never been complete, nor has it ever been unproblematic. Many of the characteristics of Maoism – e.g. 'industrialisation without urbanisation', the co-constitutive roles between state-owned enterprises (hereafter SOEs) and rural enterprises, common ownership of means of production, state-dominated financial system, etc. – extended at the very least into the early 1990s. At this point it is unclear if foundational change across the national scale is on the cards, or whether change is delimited to the selected territories in order to retain core regulatory functions (i.e. the concentration of control over land, labour and money).

Two distinct exceptions could be discerned, however. The first is the nationwide implementation of rural household-based production in the first half of the 1980s (a project that, to be sure, was not novel to begin with: Mao vehemently objected repeated calls for its implementation in the 1950s and early 1960s). By producing a huge surplus of labour power in the rural hinterland, this institution literally impelled this surplus to seek wage labour, and opportunities for wage employment were largely located in coastal city-regions privileged for capital allocation. Meso-scale policy shifts in the rural hinterland thereby became a distinct condition of possibility for the coastal-oriented 'industrialisation through urbanisation' spatial strategy in the 1990s and 2000s (pace Huang 2008; Coase and Wang 2012; cf. Table 1.3). The second exception is the overhaul of the urban–rural 'dual structure' in Chongqing (ref. Chapter 7. In spite of strong objections to these policies, the fact that the reform policies instituted by the Bo regime remain in place after his tenure in March 2012 establishes an important path-changing precedent. All this said, the retention of the nationalised land management system and *hukou* institution – key policies launched during the

1958 'socialist high tide' – across the country offers a reminder that their under-lying regulatory logics are still of relevance to the reproduction of Chinese statehood.

New Research Directions

The narrative presented in *On Shifting Foundations* leads to new research ques-tions on the connections between state rescaling, policy experimentation, and institutional path-dependency. Specifically, the book invites further conversations with social scientists on the constitutive roles of state rescaling and policy exper-imentation in the socioeconomic 'transition' of contemporary China. At the macro level, it highlights the need for a broader comparative study that examines how the Chinese state apparatus managed to build on its Maoist past without experiencing the massive socio-political instability that emerged in the former Soviet bloc. Indeed, the 'relative' political stability of 'transition' in China – the term 'relative' is used deliberately in light of the 1989 Tiananmen riots, arguably the most serious challenge to CPC rule since 1949 – could be attributed to the Chinese central government's adroit utilisation of geographically-targeted policy experimentation. In so doing, the regulatory logics of Mao-era institutions would not be suddenly destabilised, while still providing the CPC with the room to experiment with more flexible modes of socioeconomic regulation. Political sta-bility ensued because decentralisation became an aspect of centralised gover-nance. Understanding how this phenomenon unfolds across all 'nationally strategic' locations (19 at the time of writing) would offer incisive insights into the structural coherence of the Chinese political economy.

In relation to research on national regulatory changes in Chinese studies, the book calls attention to how the processes within the strategically and spatially selected policy experimentation open up new developmental pathways that are not necessarily charted, in advance, as part of some preconceived national plan. At one level, this ability to respond strategically through spatial reconfiguration arguably reproduces adaptive governance in the central government. This is espe-cially when the geographically-variegated effects of previous developmental policies call for nuanced and context-specific approaches to experimental reforms. Yet it is important to emphasise that this renewed adaptive capacity is in part contingent on the repurposing and re-tasking of regulatory capacities inher-ited from earlier regimes. The past, in other words, was never jettisoned in toto. In this respect, blunt portrayals of the post-Mao era as emblematic of 'decentral-ised governance' (in relation to centralised governance) or 'capitalism' (in rela-tion to 'socialism') will not adequately explain the transitional present. As mentioned earlier, this is a transition that is at once 'hybrid', 'coherent' and yet 'without a theory' (Shirk 1993; Naughton 1995; Rawski 1995; Zhu 2007). And as the case studies have shown, new dimensions to the 'transition' have been added because political actors seek to launch experimental reforms in selected territories designated as 'nationally strategic'. Where, then, will this 'transition' take the central government? Or, to view this from another angle, is sustained central governance now largely predicated on the roll out of new spaces of policy experimentation?

These questions connect to and build on the re-evaluation of the continuities and changes in regulatory logics during the two periods. As discussed in Chapter 2, the regulatory emphasis of the Mao era was never about economic–geographical egalitarianism (in terms of output or income); it was about political mobilisation, about economic nationalism, and about prioritising revolutionary consciousness over – or in spite of – deficiencies of economic projects. In Harvey's (1996) parlance, the primary 'permanence' Mao sought was 'permanent revolution', and he believed this 'permanence' would drive China to economic greatness. What Deng and his successors did was to reinstate household-based capital accumulation; the 'permanences' they wanted were (and still are) subnational economic geographies that could compete within the global system of capitalism (cf. Breslin 2000; Zhang and Peck 2016).

Yet whether national economic planning could be achieved in the face of increasing regulatory constraints (as highlighted in Table 1.1, Chapter 1) and global neoliberalising pressures remains an open empirical question. As the two case studies indicate (Chapters 4–7), the formation of 'nationally strategic new areas' at once reflects the lack of a centralised 'national agenda' and the strength of the post-1949 unitary political system in which the central government retains absolute power over the subnational governments. It reflects specifically a *positive* feedback loop between decentralised policy experiments, economic–geographical competitiveness, and central political power. To ascertain further the stability of this feedback loop, policies instituted in other 'nationally strategic new areas' have to be examined. This would test further the claim that the post-Mao period was not marked by a simple turn to decentralised governance. Rather, if scholars were to develop the challenging proposition by Jin Yan, Qin Hui, and Wen Tiejun that central economic planning is only blossoming in the contemporary juncture (ref. Chapter 2), the crucial question is whether and how this planning could work in relation to neoliberal regulatory logics that are currently hegemonic at the global scale. As this book has shown, the CPC's ability to generate the economic 'miracle' of the past 30 years is arguably contingent on the dynamic central-local engagement to launch 'nationally strategic' policy experimentation in targeted geographies. How 'central planning' in China is defined in the future is thus an important focal point for future research: it potentially generates new insights on whether the CPC's experimental developmental approach could produce place-specific pathways for growth in ways that do not simultaneously undermine the institutional foundations that enable it to stay in power forever. *On Shifting Foundations* offers a starting point for this exciting exploration.

References

21st Century Business Herald (2010) Zhuanfang Chongqing shizhang Huang Qifan: Guozi yu jinrong ladong Chongqing fazhan [Interview with Chongqing mayor Huang Qifan: state capital and finance pulls along Chongqing development], 17 April.

21st Century Business Herald (2012) Qianhai xiangmu huo shoubi yinhang daikuan, Guokai hang chihuan 2 yi yuan [Qianhai projects obtain first batch of bank loans, CDB advances 200 million], 3 August.

21st Century Business Herald (2013a) Chongqing shi guoziwei: badatou xianxing, guoqi, minqi ronghe fazhan [Chongqing SASAC: private and state-owned firms integrated development within eight major investments' experiment], 9 March.

21st Century Business Herald (2013b) Zhuanfang Xiao Jincheng: 'Xinqu zhanlüe' jiangcheng wangshi? [Exclusive interview with Xiao Jincheng: 'New Area strategy' a matter of the past?], 27 December.

Ahlers, A. and Schubert, G. (2015) Effective policy implementation in China's local state. *Modern China*, 41(4), 372–405.

Anagnost, A. (1993) Nationscape: movement in the field of vision. *Positions*, 1(3), 585–606

Andreas, J. (2010) A shanghai model? On capitalism with Chinese characteristics, *New Left Review*, 65, 63–85.

Asahi Shimbun (2013) Hu Shuli: China still has not compiled a common dream. 18 April.

Asia Times (2009) Power struggle behind revival of Maoism. 24 November. Retrieved on 22 March 2014 at: http://www.atimes.com/atimes/China/KK24Ad01.html.

Bao, T. (2008) 'Two Faces' of Deng Xiaoping, Translated and published by Radio Free Asia. Accessed on 10 June 2015 at: http://www.rfa.org/english/news/china/baotong-1229 2008165015.html.

Bayırbağ, M. K. (2013) Continuity and change in public policy: redistribution, exclusion and state rescaling in Turkey. *International Journal of Urban and Regional Research*, 37(4), 1123–1146.

Bedeski, R. E. (1975) The evolution of the modern state in China: nationalist and communist continuities. *World Politics*, 27(4), 541–568

Bernstein, T. P. and Li, H. Y. (eds.) (2010) *China Learns from the Soviet Union, 1949–Present*. Plymouth: Lexington Books.

Bramall, C. (2007) *The Industrialization of Rural China*. New York: Oxford University Press.

Bray, D. (2005) *Social Space and Governance in Urban China: The Danwei System from Origins to Reform*. Stanford: Stanford University Press.

Breathnach, P. (2010) From spatial keynesianism to post-fordist neoliberalism: emerging contradictions in the spatiality of the Irish State. *Antipode*, 42, 1180–1199.

Brenner, N. (2004) *New State Spaces: Urban Governance and the Rescaling of Statehood*. Oxford: Oxford University Press.

Brenner, N. (2009) Open questions on state rescaling. *Cambridge Journal of Regions, Economy and Society*, 2(1), 123–139.

Breslin, S. (2000) Decentralisation, globalisation and china's partial re-engagement with the global economy. *New Political Economy*, 5(2), 205–226.

Breslin, S. (2007) The political economy of development in China: political agendas and economic realities. *Development*, 50(3), 3–10.

Breslin, S. (2014) Financial transitions in the PRC: banking on the state? *Third World Quarterly*, 35(6), 996–1013.

Brødsgaard, K. E. (2012) Politics and business group formation in China: the party in control? *The China Quarterly*, 211, 624–648.

Brown, J. (2014) *City Versus Countryside in Mao's China: Negotiating the Divide*. Cambridge: CUP.

Burns, J. P. (1999) The people's republic of China at 50: national political reform. *The China Quarterly*, 159, 580–594.

Burns, J. P. (2003) "Downsizing" the Chinese state: government retrenchment in the 1990s. *The China Quarterly*, 175, 775–802.

Cai, H. and Treisman, D. (2006) Did government decentralization cause China's economic miracle?. *World Politics*, 58(4), 505–535.

Caijing (2010a) Chongqing hugai yuejin. [Leap in hukou reforms in Chongqing] 30 August.

Caijing (2010b) Huang Qifan tan Chongqing hugai. [Huang Qifan talks about hukou reforms in Chongqing] 30 August.

Caijing (2013) Huang Qifan jiexi Chongqing jingji [Huang Qifan explains the Chongqing economy]. 13 March.

Caijing (2014) Chongqing guoqi gaige zhongdian fazhan hunhe jingji [Chongqing SOE reforms to emphasize mixed economy], 20 January. Accessed on 24 April 2014 at: http://economy.caijing.com.cn/2014-01-20/113836922.html

Caixin (2012) China's Gini Index at 0.61, University Report Says. Caixin Online 11 December. Accessed on 12 December 2012 at: http://english.caixin.com/2012-12-10/100470648.html

Caixin (2015) How Beijing intervened to save China's stocks. 16 July. Accessed on 18 July 2015 at: http://english.caixin.com/2015-07-16/100829521.html

CCCPC Party Literature Research Office (1998) *Jianguo yilai Mao Zedong Wengao*, Vol. 13. Beijing: Zhongyang Wenxian Chubanshe.

CCTV (2012) Full text of Hu Jintao's report at 18th Party Congress. 8 Nov. Official English full text released 18 November. Accessed on 22 November 2012 at: http://english.cntv.cn/20121118/100129.shtml

Chan, K. W. (1994) *Cities with Invisible Walls: Reinterpreting Urbanization in Post-1949 China.* Hong Kong: Oxford University Press.

Chan, K. W. (2010a) A China paradox: migrant labor shortage amidst rural labor supply abundance. *Eurasian Geography and Economics,* 51(4), 513–530.

Chan, K. W. (2010b) The global financial crisis and migrant workers in China: 'there is no future as a labourer; returning to the village has no meaning'. *International Journal of Urban and Regional Research,* 34(3), 659–677.

Chan, K. W. and Buckingham, W. (2008) Is China abolishing the hukou system? *The China Quarterly,* 195, 582–606.

Chang, J. K. (1969[2010]) *Industrial Development in Pre-Communist China: 1912–1949.* London: Transaction Publishers.

Chang, J. and Halliday, J. (2007) *Mao: The Unknown Story.* London: Random House.

Chen, Y. (1995) *Chen Yun Wen Xuan,* Vol. 2. Beijing: Renmin Chubanshe.

Chen, Y. (2000) *Chen Yun Nianpu (Xiajuan).* Beijing: Zhongyang Wenxian Chubanshe.

Chen, T.J. and Ku, Y.H. (2014). Indigenous innovation vs. Teng-Long Huan-Niao: policy conflicts in the development of China's flat panel industry. *Industrial and Corporate Change,* 23(6), 1445–1467.

Chien, S. S. (2010) Economic freedom and political control in post-Mao China: a perspective of upward accountability and asymmetric decentralization. *Asian Journal of Political Science,* 18(1), 69–89.

Chien, S. S. and Gordon, I. (2008) Territorial competition in China and the West. *Regional Studies,* 42(1), 1–18.

China Business Times (2012) Yang Qingyu: Shenhua gaige yingzhuzhua boruo huanjie. [Yang Qingyu: Reform deepening should focus on weak links]. 8 March.

China Daily (2010) Wen upbeat on US relations despite strains. 23 March.

China Daily (2011) Full text of Hu's speech at opening of Boao Forum. 11 April.

China Daily (2012) Income gap needs attention: Bo Xilai. 6 March.

China Daily (2013) Showtime for modern Chongqing. 7 June.

China Economic Times (2001) "'Yige Zhongguo, sige shijie' xi diqu fazhan chaju" ['One China, four worlds': analyzing regional developmental disparities]. 17 April.

China Economic Weekly (2006) Buneng rang zhuhou jingji boyi hongguan tiaokong [Cannot allow feudal economics turn macro adjustment into zero-sum game]. 11 December. Accessed on 31 May 2014 at: http://finance.people.com.cn/GB/70392/5151782.html.

China Economic Weekly (2013) Chongqing meiyou moshi, zhiyou zhihuixing gaige de shenru he tansuo. [Chongqing has no model, only an intellectually-based deepening and exploration of reforms] 18 March. Accessed on 24 March 2013 from: http://www.ceweekly.cn/html/Article/201303185203254011.html.

China Economy & Informatization (2012) Zhuanfang Huang Qifan: Chongqing buxuyao tuilu. [Interview with Huang Qifan: Chongqing does not need to retreat] 18 January.

China Entrepreneur (2013) Wu Jinglian: Shehui maodun yidao jixian, gaige buneng zaicuo. [Wu Jinglian: Social contradictions have reached limit, reforms cannot continue to blunder] 1 November.

China News (2013a) Hengqin shenhua gaige yu gaishan minsheng de pinghengdian. 8 November. Retrieved on 12 December 2013 from: http://www.chinanews.com/gn/2013/11-08/5478692.shtml.

China News (2013b) Liangjiang Xinqu: Jingji jiegou buduan youhua. [Liangjiang New Area: Economic structure enhanced]. 26 December.

China Reform (2010) Chongqing 'xin jingji zhengce' ['New economic policy' in Chongqing]. Issue 11 (November).

China Reform (2012) Hu Deping: Gei renmin yige you anquangan dehuanjing. [Hu Deping: Give people a secure environment]. 4 December. Retrieved 11 February 2014 at: http://business.sohu.com/20121204/n359454195.shtml.

China Review (2013) Cong shenme jiaodu kandai Chongqing moshi. [From what angle to look at Chongqing model] 15 January. Accessed on 29 September 2013 at: http://china-review.com/sbao.asp?id=4950&aid=31265.

China Securities Journal (2008) "Wen Jiabao zai Guangdong diaoyan, qiangdiao jiada zhi-chi zhongxiao qiye lidu", 21 July.

Chinese State Council 6th Population Census Office (2012). *Major Figures on 2010 Population Census of China* [in Mandarin]. Beijing: China Statistics Press.

Chongqing Chengbao (2010) Chongqing gongzufang guanli banfa chutai 7yue1ri qi zhengshi shixing. [Regulations for Chongqing public rental housing to roll-out on 1 July] 9 June.

Chongqing Municipal Bureau of Statistics (2013) *Chongqing StatisticalYearbook 2012*. Beijing: China Statistics Press.

Chongqing Municipal Bureau of Statistics (2017) *Chongqing StatisticalYearbook 2017* Beijing: China Statistics Press.

Chongqing Research Academy of Education Sciences (2015) *Chongqing Lishi [History of Chongqing]*. Chongqing: Southwest China Normal University Press.

Chongqing Ribao (2007) 314 zongtibushu gouhua Chongqing fazhan lantu. [314 integrated positioning a blueprint for Chongqing development]. 8 October.

Chongqing Ribao (2010a) Liangjiang Xinqu shi Zhongguo fazhan zhanlüe zhuanxing de tupokou [Liangjiang New Area is the breakthrough point for strategic readjustment in China]. 6 August.

Chongqing Ribao (2010b) Chongqing gongzufang jianshe jinzhan shunli, zhujin jinwei shichangjia 60%. [Public rental housing in Chongqing progresses smoothly, rental is 60% of market price]. 28 October.

Chongqing Ribao (2010c) Yangshi zhuanfang Huang Qifan xiangjie Chongqing gongzu-fang. 5 September.

Chongqing Ribao (2011) Liangjiang xinqu shi xibu jingji fadongji. [Liangjiang New Area is the economic motor of the west] 17 June.

Chongqing Ribao (2012a) Guofa 3 hao wenjian yinling Chongqing liebian. [National no. 3 document leads change in Chongqing] 14 January.

Chongqing Ribao (2012b) "Yang Qingyu: Chongqing bixu niuzhuan jini xishu jixu shang-sheng shitou". [Yang Qingyu: Chongqing must overturn the GINI-coefficient and con-tinue upgrading momentum] 10 March.

Chongqing Ribao (2014) Liangjiang Xinqu 2013nian shixian GDP 1650 yiyuan [Liangjiang New Area GDP reached 165 billion yuan]. 25 January.

Chongqing Shangbao (2011) 2010 nian Chongqing CPI he chengzhen jumin renjun kezhi-pei shouru gongbu. 28 January.

Chongqing Shibao (2018) Shijie 500 qiang qiye 147 jia luohu Liangjiang [147 of Fortune 500 firms are established in Liangjiang]. 22 June.

Chongqing Wanbao (2010) "Huang Qifan jieshou Yang Lan zhuanfang jiangshu Chongqing huji gaige zhengce chutai qianhou", 25 August.

Chung, J. H. (2000) *Central Control and Local Discretion in China-leadership and Implementation During Post-Mao Decollectivization.* New York: Oxford University Press.

Coase, R. and Wang, N. (2012) *How China Became Capitalist.* London: Palgrave.

Cohen, P. A. (1988) The post-Mao reforms in historical perspective. *The Journal of Asian Studies,* 47(03), 518–540.

Cox, K. R. (2009) 'Rescaling the state' in question. *Cambridge Journal of Regions, Economy and Society,* 2, 107–121.

Cox, K. R. (2010) The problem of metropolitan governance and the politics of scale. *Regional Studies,* 44, 215–227.

Dabla-Norris, M. E. (2005) Issues in intergovernmental fiscal relations in China. International Monetary Fund Working Paper, WP/05/30. Available at: https://www.imf.org/en/Publications/WP/Issues/2016/12/31/Issues-in-Intergovernmental-Fiscal-Relations-in-China-17893.

Davoudi, S. (2009) Scalar tensions in the governance of waste: the resilience of state spatial Keynesianism. *Journal of Environmental Planning and Management,* 52(2), 137–156.

Day, A. (2013) *The Peasant in Postsocialist China: History, Politics, and Capitalism.* Cambridge: Cambridge University Press.

de Jong, M. (2013) China's art of institutional bricolage: selectiveness and gradualism in the policy transfer style of a nation. *Policy and Society,* 32(2), 89–101.

Deng, X. (1994a) *Deng Xiaoping Wenxuan [Deng Xiaoping's Selected Works],* Vol. 1. Beijing: Renmin Chubanshe.

Deng, X. (1994b) *Deng Xiaoping Wenxuan [Deng Xiaoping's Selected Works],* Vol. 2. Beijing: Renmin Chubanshe.

Deng, L. (1998) *Mao Zedong du shehui zhuyi zhengzhi jingjixue pizhu he tanhua [Mao Zedong's Commandments and Speeches on Reading Socialist Political Economy].* Beijing: PRC Historical Society.

Development Research Center of the State Council (DRCSC). (2004) A report on local protectionism in China. *References for Economic Research,* 18, 31–38.

Dickson, B. (2008) *Wealth into Power: The Communist Party's Embrace of China's Private Sector.* Cambridge: Cambridge University Press.

Dikötter, F. (2010) *Mao's Great Famine: The History of China's Most Devastating Catastrophe,* 1958–1962. London: Bloomsbury Publishing.

Dikötter, F. (2016) *The Cultural Revolution: A People's History,* 1962–1976. London: Bloomsbury Publishing.

Diyi Caijing Ribao (2012) Wunian nitou 1810 yi, Chongqing Liangjiang Xinqu duibiao Pudong [Plans to invest 181 billion yuan in 5 years, Chongqing Liangjiang New Area aims at Pudong]. 22 February. Retrieved on 19 August 2018 at: https://finance.qq.com/a/20120222/000851.htm.

Donnithorne, A. (1972) China's cellular economy: some economic trends since the cultural revolution. *China Quarterly*, 52, 605–619.

Duara, P. (1991). *Culture, Power, and the State: rural North China*, 1900–1942. Stanford: Stanford University Press.

Dunford, M. and Li, L. (2010) Chinese spatial inequalities and spatial policies. *Geography Compass*, 4(8), 1039–1054.

DushiKuaibao. (2011). Chongqing shizhang Huang Qifan lianghui jieshou zhuanfang, huanying canyu Liangjiang Xinqu jianshe. [Chongqing mayor Huang Qifan welcomes participation in the construction of Liangjiang during interview at 'Two Meetings']. 7 March.

Economy, E. C. (2011) *The River Runs Black: The Environmental Challenge to China's Future*. Ithaca, NY: Cornell University Press.

Economy & Nation Weekly (2011, January 10) Bo Xilai + Huang Qifan: Chongqing tudi huji gaige. [Bo Xilai + Huang Qifan: reforms of huji and land in Chongqing].

ENN Weekly (2014) Zhuanfang Huang Qifan: Chongqing guozi gaige zaichufa [Interview with Huang Qifan: Relaunch of state capital reforms in Chongqing]. 21 January.

Fan, C. C. (1995) Of belts and ladders: state policy and uneven regional development in post-Mao China. *Annals of the Association of American Geographers*, 85(3), 421–449.

Fan, C. C. (2002) The elite, the natives, and the outsiders: migration and labor market segmentation in urban China. *Annals of the Association of American Geographers*, 92(1), 103–124.

Fan, C. C. (2008) "Migration, Hukou, and the Chinese City," in *China Urbanizes: Consequences, Strategies, and Policies*, edited by Shahid Yusuf and Anthony Saich Washington, D.C.: The World Bank. pp. 65–90.

Financial Times (2012). Difang zhengfu jingzheng yu zhaiwufengxian. [Local government competition and debt-related risks] Chinese edition.9 March. Accessed on 29 April 2012 at: http://www.ftchinese.com/story/001043561.

Financial Times (2017) China rejects Singapore model for state-owned enterprise reform. 20 July.

Fitzgerald, J. (1995) The nationless state: the search for a nation in modern Chinese nationalism. *The Australian Journal of Chinese Affairs*, (33), 75–104.

Florini, A., Lai, H., and Tan, Y. (2012) *China Experiments: From Local Innovations to National Reform*. Washington, DC: Brookings Institution Press.

Fong, V. (2011) *Paradise Redefined: Transnational Chinese Students and the Quest for Flexible Citizenship in the Developing World*. Stanford: Stanford University Press.

FSDC (2013) Proposals to Advance the Developmment of Hong Kong as an Offshore Renminbi Centre. FSDC Research Paper No. 03, 1–8.

Global Post (2012) In China, hot money crosses borders. 19 April. Available at: https://www.pri.org/stories/2012-04-19/china-hot-money-crosses-borders.

Göbel, C. (2011) Uneven policy implementation in rural China. *The China Journal*, (65), 53–76.

Goldstein, S. M. (1995) China in transition: the political foundations of incremental reform. *The China Quarterly*, 144, 1105–1131.

Greenhalgh, S. and Winckler, E. A. (2005) *Governing China's Population: From Leninist to Neoliberal Biopolitics*. Stanford: Stanford University Press.

Gries, P. H. (2004) *China's New Nationalism: Pride, Politics and Diplomacy*. Berkeley and London: University of California Press.

Guangdong Bureau of Statistics (2008) *Guangdong Statistical Yearbook 2007*. Beijing: China Statistics Press.

Guangdong Bureau of Statistics (2013) *Guangdong Statistical Yearbook 2012*. Beijing: China Statistics Press.

Guangdong Bureau of Statistics (2014) *Guangdong Statistical Yearbook 2013*. Beijing: China Statistics Press.

Guangzhou Daily (2008a) Jintian bu jiji tiaozheng chanye jiegou, mingtian jiuyao bei chanye jiegou tiaozheng. 27 March.

Guangzhou Daily (2008b) Wang Yang: Miandui jingji kunnan yao you xinxin, zhengfu bujiu luohou shengchanli, 14 November.

Guangzhou Daily (2009) Hu Jintao dao Guangdong kaocha, qiangdiao yao jiandingbuyi de tiao jiegou, jiaotashidi de chu zhuanxing. [Hu Jintao emphasizes resolute and grounded structural adjustment during inspection of Guangdong]. 30 December.

Guo, F. and Huang, Y. S. (2010) Does "hot money" drive China's real estate and stock markets? *International Review of Economics & Finance*, 19(3), 452–466.

Harrison, J. (2010) Networks of connectivity, territorial fragmentation, uneven development: the new politics of city-regionalism. *Political Geography*, 29(1), 17–27.

Harrison, J. (2012) Life after regions? The evolution of city-regionalism in England. *Regional Studies*, 46(9), 1243–1259.

Harvey, D. (1985) *The Urbanization of Capital*. Oxford: Blackwell.

Harvey, D. (1996) *Justice, Nature and the Geography of Difference*. Oxford: Blackwell.

Harvey, D. (2010) *The Enigma of Capital*. New York: Oxford University Press.

Hayter, R. and Barnes, T. (1990) Innis' staple theory, exports, and recession: British Columbia, 1981–1986. *Economic Geography*, 66(2), 156–173.

Hayter, R. and Barnes, T. (2001) Canada's resource economy. *The Canadian Geographer*, 45(1), 36–41.

He, X. (2010) *Diquan de luoji: Zhongguo nongcun tudi zhidu xiang hechu qu [Logics of Land Rights:Where Is the Rural Land Institution in China Heading?]*. Beijing: Zhongguo Zhengfa Daxue Chubanshe.

He, S. and Wu, F. (2009) China's emerging neoliberal urbanism: perspectives from urban redevelopment. *Antipode*, 41(2), 282–304.

He, C. and Yang, R. (2016) Determinants of firm failure: empirical evidence from China. *Growth and Change*, 47(1), 72–92.

Heilmann, S. (2008) Policy experimentation in China's economic rise. *Studies in Comparative International Development*, 43(1), 1–26.

Heilmann, S. (2009) Maximum tinkering under uncertainty: unorthodox lessons from China. *Modern China*, 35(4), 450–462.

Heilmann, S. and Perry, E. (eds.) (2011) *Mao's Invisible Hand: The Political Foundations of Adaptive Governance in China*. Cambridge, MA: Harvard University Press.

Hinton, W. (1990) *The Great Reversal: The Privatization of China*. New York: Monthly Review Press.

Horesh, N. and Lim, K. F. (2017) China: an East Asian alternative to neoliberalism? *The Pacific Review*, 30(4), 425–442.

Howell, J. (2006) Reflections on the Chinese state. *Development and Change*, 37(2), 273–297.

Hsing, Y. T. (2010) *The Great Urban Transformation: Politics of Land and Property in China*. Oxford: Oxford University Press.

Hsueh, R. (2011) *China's Regulatory State: A New Strategy for Globalization*. Ithaca, NY: Cornell University Press.

Huang, Y. (1996) Central-local relations in China during the reform era: the economic and institutional dimensions. *World Development*, 24(4), 655–672.

Huang, Y. (2004) Housing markets, government behaviors, and housing choice: a case study of three cities in China. *Environment and Planning A*, 36(1), 45–68.

Huang, Y. (2008) *Capitalism with Chinese Characteristics: Entrepreneurship and the State*. New York: Cambridge University Press

Huang, B. and Chen, K. (2012) Are intergovernmental transfers in China equalizing? *China Economic Review*, 23(3), 534–551.

Huang, Y. and Yang, D. L. (1996) The political dynamics of regulatory change: speculation and regulation in the real estate sector. *Journal of Contemporary China*, 5(12), 171–185.

ifeng.com (2013) Wen Tiejun: Nongcun chengdan Zhongguo baci weiji daijia [Wen Tiejun: The price and burden of China's eight crises for the countryside]. Full transcript of speech in Peking University. 11 January. Available online at: http://blog.sina.com.cn/s/blog_b565cf140101umic.html.

Jackson, S. (2011) A tale of two state-builders: kuomintang and communist compared. *Columbia East Asia Review*, 4, 76–88.

Jessop, B. (1993) Towards a Schumpeterian workfare state? Preliminary remarks on post-Fordist political economy. *Studies in Political Economy*, 40, 7–39.

Jessop, B. (2001) What follows Fordism? On the periodization of capitalism and its regulation. In: R. Albritton and M. Itoh (eds.), *Phases of Capitalist Development: Booms, Crises, and Globalization*, pp. 282–299. Basingstoke: Palgrave.

Jessop, B. (2002) *The Future of the Capitalist State*. Cambridge: Polity.

Jessop, B. (2016) *The State: Past, Present, Future*. Malden, MA: Polity.

Jessop, B., Brenner, N., and Jones, M. 2008 Theorizing sociospatial relations. *Environment and Planning D: Society and Space*, 26(3), 389–401.

Jessop, B. and Sum, N. L. (2006) *Beyond the Regulation Approach: Putting Capitalist Economies in their Place*. Cheltenham: Edward Elgar Publishing.

Jiang, Z. (1999) Quandang quanshehui jinyibu dongyuan qilai, shouqu baqi fupin gongjian juezhan jieduan de shengli. [To intensify the mobilisation of the whole Party and society and obtain victory in the fortification stage of the '87' poverty alleviation program]. 9 June. Available at: http://cpc.people.com.cn/GB/64184/64186/66689/4494535.html.

Jiang, Z. (2006) *Jiang Zemin Wenxuan* Vol. 2. Beijing: Renmin Chubanshe.

Jonas, A. (2013) City-regionalism as a contingent 'geopolitics of capitalism'. *Geopolitics*, 18, 284–98.

Jones, M. (2001) The rise of the regional state in economic governance: 'partnerships for prosperity' or new scales of state power?. *Environment and Planning A*, 33, 1185–1211.

Jones, P. P. and Poleman, T. T. (1962) Communes and the agricultural crisis in communist China. *Population*, 600(500), 1–20.

Kanbur, R. and Zhang, X. (1999) Which regional inequality? The evolution of rural–urban and inland–coastal inequality in China from 1983 to 1995. *Journal of Comparative Economics*, 27(4), 686–701.

Kanbur, R. and Zhang, X. (2005) Fifty years of regional inequality in China: a journey through central planning, reform, and openness. *Review of Development Economics* 9(1), 87–106.

Korean Joongang Daily (2012) Xi Jinping shidai, zhongguo jiang xianqi shichanghua, fazhihua gaige de fenglang. [A marketizing and rule of law wave will arise in China during the Xi Jinping era]. 20 April.

Kuhn, P. A. (2002) *Origins of the Modern Chinese state*. Stanford: Stanford University Press.

Lan, J., Kakinaka, M. and Huang, X. (2012) Foreign direct investment, human capital and environmental pollution in China. *Environmental and Resource Economics*, 51(2), 255–275.

Lawrence, S. and Martin, M. F. (2013) Understanding China's political system. *CRS Report for Congress*. 20 March.

LeBlanc, B. (2014) Illicit Trade Invoicing Fuels China's Currency, Housing Speculation. Op-ed for *Global Financial Integrity*. Available here: https://www.gfintegrity.org/press-release/illicit-trade-invoicing-fuels-chinas-currency-housing-speculation.

Lee, P. K. (1998) Local economic protectionism in China's economic reform. *Development Policy Review*, 16, 281–303.

Lefebvre, H. (1991) *The Production of Space*. Oxford: Blackwell.

Ley, D. and Hutton, T. (1987) Vancouver's corporate complex and producer services sector: linkages and divergence within a provincial staple economy. *Regional Studies*, 21(5), 413–424.

Levine, R. and Zervos, S. (1998a) Stock markets, banks, and economic growth. *American Economic Review*, 88, 537–558.

Levine, R. and Zervos, S. (1998b) Capital control liberalization and stock market development. *World Development*, 26(7), 1169–1183.

Li, P. (1988) 1988 nian guowuyuan zhengfu gongzuo baogao [1988 State Council Work Report], 25 March. Accessed on 6 May 2014 at: http://www.gov.cn/test/2006-02/16/content_200865.htm.

Li, K. (1991) Lun woguo jingji de sanyuan jiegou [On our country's 3-tiered structure]. *Zhongguo Shehui Kexue*, 3(1991), 65–82.

Li, L. C. (1997) Towards a non-zero-sum interactive framework of spatial politics: the case of centre–province in contemporary China. *Political Studies*, 45(1), 49–65.

Li, P. (2003) *Zong Zhi Hui Hongtu: Li Peng San Xia Riji [Li Peng's diary from the Three Gorges]*. Beijing: China Three Gorges Publishing House.

Li, H. -Y. (2006) *Mao and the Economic Stalinization of China, 1949–1953*. Lanham, MD: Rowman and Littlefield.

Li, C. (2008) "Zhongguo Nongmin de zizhuxing yu Zhongguo de zizhuxing", 21 August. Accessed online on 8 August 2013 at: http://www.caogen.com/blog/infor_detail.aspx?articleId=9838&Page=1.

Li, C. (2016) Historical observations regarding the large-scale establishment of rural public canteens in Hebei province. In H. Li and T. Dubois (Eds.) *Agricultural Reform and Rural Transformation in China since 1949*. Leiden & Boston: Brill, pp. 115–132.

Li, L. (2010) *Breaking Through: The Birth of China's Opening-up Policy*. Oxford: Oxford University Press.

Li, J. and Hsu, S. (2012) Shadow banking in China. Working paper. Accessed on 31 May 2015 at: http://mpra.ub.uni-muenchen.de/39441.

Li, C. and Gibson, J. (2013) Rising regional inequality in China: fact or artifact?. *World Development*, 47, 16–29.

Li, H. and Rozelle, S. (2003) Privatizing rural China: insider privatization, innovative contracts and the performance of township enterprises. *The China Quarterly*, 176, 981–1005.

Li, H. and Rozelle, S. (2004) Insider privatization with a tail: the screening contract and performance of privatized firms in rural China. *Journal of Development Economics*, 75(1), 1–26.

Li, Y., and Wu, F. (2012) The transformation of regional governance in China: the rescaling of statehood. *Progress in Planning*, 78(2), 55–99.

Li, Y. and Wu, F. (2013) The emergence of centrally initiated regional plan in China: a case study of Yangtze River Delta Regional Plan. *Habitat International*, 39, 137–147.

Li, Y. and Wu, F. (2018) Understanding city-regionalism in China: regional cooperation in the Yangtze River Delta. *Regional Studies*, 52(3), 313–324.

Li, Y., Wu, F., and Hay, I. (2015) City-region integration policies and their incongruous outcomes: the case of Shantou-Chaozhou-Jieyang city-region in east Guangdong Province, China. *Habitat International*, 46, 214–222.

Li, Z., Xu, J., and Yeh, A. G. (2014) State rescaling and the making of city-regions in the Pearl River Delta, China. *Environment and Planning C: Government and Policy*, 32, 129–143.

Lim, K. F. (2010) On China's growing geo-economic influence and the evolution of variegated capitalism. *Geoforum*, 41, 677–688.

Lim, K. F. (2012) What you see is (not) what you get? The Taiwan question, geo-economic realities, and the "China Threat" imaginary. *Antipode*, 44(4), 1348–1373.

Lim, K. F. (2014) 'Socialism with Chinese characteristics': Uneven development, variegated neoliberalization and the dialectical differentiation of state spatiality. *Progress in Human Geography*, 38, 221–247.

Lim, K. F. (2017) Researching state rescaling in China: methodological reflections. *Area Development and Policy*, online version.

Lin, G.C.S. (1997) *Red Capitalism in South China: Growth and Development of the Pearl River Delta*. Vancouver: UBC Press.

Lin, G. C. S. (1999) State policy and spatial restructuring in post-reform China, 1978–1995. *International Journal of Urban and Regional Research*, 23(4), 670–696.

Lin, G. C. S. (2007) Chinese urbanism in question: State, society, and the reproduction of urban spaces. *Urban Geography*, 28, 7–29.

Lin, G. C. S. (2009) *Developing China: Land, Politics, and Social Conditions*. London & New York, NY: Routledge.

Lin, N. (2011) Capitalism in China: a centrally managed capitalism (CMC) and its future. *Management and Organization Review*, 7(1), 63–96.

Lin, L. W. and Milhaupt, C. J. (2013) We are the (national) champions: understanding the mechanisms of state capitalism in China. *Stanford Law Review*, 65, 697–760.

Liu, H. (2006) Changing regional rural inequality in China 1980–2002. *Area*, 38(4), 377–389.

Liu, W., Jin, F., Liu, Y., Liu H., Zhang, W. and Lu, D. (2012) *2011 Zhongguo Quyu Fazhan Baogao [2011 report on regional report in China]*. Beijing: Shangwu Yinshuguan.

Lobao, L., Martin, R., and Rodríguez-Pose, A. (2009) Editorial: rescaling the state: new modes of institutional–territorial organization. *Cambridge Journal of Regions, Economy and Society*, 2, 3–12.

Lu, L. and Wei, Y.D. (2007) 'Domesticating Globalisation, New Economic Spaces and Regional Polarisation in Guangdong Province, China'. *Tijdschrift Voor Economische en Sociale Geografie*, 98(2), 225–244.

Lüthi, L. M. (2010) *The Sino-Soviet Split: Cold War in the Communist World*. Princeton, NJ: Princeton University Press.

Lyons, T. P. (1991) Interprovincial disparities in China: output and consumption, 1952–1987. *Economic Development and Cultural Change*, 39(3), 471–506.

Ma, L. J. (2005) Urban administrative restructuring, changing scale relations and local economic development in China. *Political Geography*, 24, 477–497.

MacLeavy, J. and Harrison, J. (2010) New state spatialities: perspectives on state, space, and scalar geographies. *Antipode*, 42(5), 1037–1046.

Man, J. Y. (2011) *Affordable housing in China*. Land Lines (Lincoln Institute of Land Policy). January.

Mao, T. (1977) *A Critique of Soviet Economics*. Translated by Moss Roberts. New York and London: Monthly Review Press.

Mao, Z. (1991) *Mao Zedong Xuanji*, Vol. 4. Beijing: Renmin Chubanshe.

Mao, Z. (1992) *Jianguo yilai Mao Zedong wengao*, Vol. 7. Beijing: Zhongyang Wenxian Chubanshe.

Mao, Z. (1999a) *Mao Zedong Wenji*, Vol. 7. Beijing: Renmin Chubanshe.

Mao, Z. (1999b) *Mao Zedong Wenji*, Vol. 8. Beijing: Renmin Chubanshe.

Mao, Y. (2013) *Zhongguoren de jiaolü cong nalilai [What are the origins of the Chinese people's worries?]* Beijing: Qunyan Press.

Maoist documentation Project: https://www.marxists.org/reference/archive/mao/selected-works/volume-8/mswv8_65.htm.

Martin, R. (2010) Roepke lecture in economic geography – rethinking regional path dependence: beyond lock in to evolution. *Economic Geography*, 86(1), 1–27.

Martin, R. and Sunley, P. (2006) Path dependence and regional economic evolution. *Journal of Economic Geography*, 6(4), 395–437.

Marton, A. M. and Wu, W. (2006) Spaces of globalisation: institutional reforms and spatial economic development in the Pudong new area, Shanghai. *Habitat International*, 30(2), 213–229.

Massey, D. ([1984] 1995) *Spatial Divisions of Labour*. London: Routledge.

Massey, D. (2005) *For Space*. London: Sage.

McKinnon, R. I. (2013) *The Unloved Dollar Standard: from Bretton Woods to the Rise of China*. New York: Oxford University Press.

McGregor, R. (2012) *The Party: The Secret World of China's Communist Rulers*, Revised ed. London: Penguin.

Meisner, M. (1999). *Mao's China and After: A History of the People's Republic*. New York: The Free Press.

Mezzadra, S. and Neilson, B. (2013). *Border as Method, or, the Multiplication of Labor*. Durham and London: Duke University Press.

Minsky, H. P. (1986) *Stabilizing An Unstable Economy*. New Haven: Yale University Press.

Montinola, G., Qian, Y., and Weingast, B. R. (1995) Federalism, Chinese style, *World Politics*, 48(1), 50–81.

Nanfang Ribao (2013) Wen Tiejun: Chengzhenhua shi yingdui weiji de xianshou [Wen Tiejun: Urbanization is a method of crisis preemption]. 23 June.

Nanfeng Chuang (2014) Xiangcun zhili yanjiu zhuanjia: Chengxiang eryuanjiegou tizhi shi dui nongmin de boduo [Village governance expert: urban–rural dual structure is exploitative of peasants]. 14 January.

Nanfang Dushibao (2008) Wang Yang huiying 'bujiu luohou shengchanli': zou ziji delu, rang bieren yilun quba. 21 November.

Nanfang Dushibao (2012a) Guangdong sannian liangdu 'daobichao' [2 waves of closures in Guangdong within 3 years], 2 April.

Nanfang Dushibao (2012b) Zhu Xiaodan lianghui toulu Hu Jintao zhan Guangdong zhuanxing 'renshi qingxing' 8 March.

Nanfang Dushibao (2013a) Qianhai shouren jüzhang, sannian buxunchang lu [Unusual road for first Qianhai Bureau chief in last 3 years]. 14 August.

Nanfang Dushibao (2013b) Dayu xiaogan, xiaoyu dagan, buxiayu wanmingdegan [We do whatever it takes]. 15 August.

Nanfang Zhoumo (2009) "Guangdong zai tanlu: Jingji xinzheng 500 tian". [Guangdong explores new path again: first 500 days of new economic governance]. 13 August.

Nanfang Zhoumo. (2010). Chongqing gongzufang: kailuo, mengjin. 26 August. [Chongqing public rental housing: roaring start, strong progress].

Nathan, A. J. (2003) Authoritarian resilience. *Journal of Democracy*, 14(1), 6–17.

National Bureau of Statistics (1992) *China Statistical Yearbook 1992*. Beijing: China Statistics Press.

National Bureau of Statistics (2008) *China Statistical Yearbook 2007*. Beijing: China Statistics Press.

National Bureau of Statistics (2013) *China Statistical Yearbook 2013*. Beijing: China Statistics Press.

National Bureau of Statistics (2014) *China Statistical Yearbook 2013*. Beijing: China Statistics Press.

National Bureau of Statistics (2017) *China Statistical Yearbook 2017*. Beijing: China Statistics Press.

National Development and Reform Commission (NDRC) (2011) Zhonggong Zhongyang guanyu zhiding guomin jingji he shehui fazhan di shierge wunian guihua de jianyi. [Suggestions on the 12th Five-Year Plan by the Communist Party of China] Available at: http://www.ndrc.gov.cn/125gh.pdf.

Naughton, B. (1988) The third front: defence industrialization in the Chinese interior. *The China Quarterly*, 115, 351–386.

Naughton, B. (1994) Chinese institutional innovation and privatization from below. *The American Economic Review*, 84(2), 266–270.

Naughton, B. (1995) *Growing Out of the Plan: Chinese Economic Reform, 1978–1993*. Cambridge: Cambridge University Press.

Naughton, B. (2008) A political economy of China's economic transition. China's great economic transformation. In: L. Brandt and T. G. Rawski (eds.), *China's Great Economic Transformation*, pp. 91–135. Cambridge: Cambridge University Press.

Naughton, B. and Yang, D. L. (eds), (2004) *Holding China Together: Diversity and National Integration in the post-Deng Era*. Cambridge: Cambridge University Press.

New York Times (2011). China's scary housing bubble. 14 April.

New York Times (2013) China's leader embraces Mao as he tightens grip on country. August 16.

Ngo, T.-W., Yin, C. and Tang, Z. (2017) Scalar restructuring of the Chinese state: the subnational politics of development zones. *Environment and Planning C*, 35(1), 57–75.

Oi, J. C. (1992) Fiscal Reform and the Economic Foundations of Local State Corporatism in China. *World Politics* 45: 99–126.

Oi J. C. (1998) The evolution of local state corporatism. In: Andrew Walder (ed.) *Zouping in Transition: The Process of Reform in Rural North China*, pp. 35–61. Cambridge: Harvard University Press.

Oi, J. C. (1999) *Rural China Takes Off: Institutional Foundations of Economic Reform*. Berkeley: University of California Press.

Olesen, K. and Richardson, T. (2012) Strategic planning in transition: contested rationalities and spatial logics in twenty-first century Danish planning experiments. *European Planning Studies*, 20(10), 1689–1706.

Ong, A. (2004) The Chinese axis: zoning technologies and variegated sovereignty. *Journal of East Asian Studies*, 4(1), 69–96.

Ong, A. (2006) *Neoliberalism as Exception: Mutations in Citizenship and Sovereignty*. Durham and London: Duke University Press.

Ong, L. H. (2012). Between developmental and clientelist states: local state-business relationships in China. *Comparative Politics*, 44(2), 191–209.

Osinsky, P. (2010) Modernization interrupted? Total war, state breakdown, and the communist conquest of China. *The Sociological Quarterly*, 51(4), 576–599.

Page, S. E. (2006) Path dependence. *Quarterly Journal of Political Science*, 1, 87–115.

Pan, J. and Li, J. (2011) *Fangdichan Lanpishu: Zhongguo Fangdichan Fazhan Baogao*. [Development Report of Real Estate Development in China]. Beijing: Zhongguo Sheke Wenxian.

Peck, J. (1998). Geographies of governance: TECs and the neo-liberalisation of 'local interests'. *Space and Polity*, 2(1), 5–31.

Peck, J. (2001) *Workfare States*. New York: Guilford.

Peck, J. (2002) Political economies of scale: fast policy, interscalar relations, and neoliberal workfare. *Economic Geography*, 78, 331–360.

Peck, J. (2003) Geography and public policy: mapping the penal state. *Progress in Human Geography*, 27(2), 222–232.

Peck, J. and Zhang, J. (2013) A variety of capitalism… with Chinese characteristics?. *Journal of Economic Geography*, 13(3), 357–396.

Pei, M. (2006) *China's Trapped Transition: The Limits of Developmental Autocracy*. Cambridge, MA: Harvard University Press.

Pei, M. (2016) *China's Crony Capitalism*. Cambridge, MA: Harvard University Press.

People's Daily (1959) Shelun: Quanguo yipanqi [Editorial: The nation as a chessboard], People's Daily, 24 February.

People's Daily (2003) Lishi shang de jintian. [Today in history] 1 August.

People's Daily (2008a) Renmin Ribao Jingji Shiping: Kuoda jiuye shandai zhongxiao qiye [Editorial: Expand employment and treat small and medium enterprises benevolently], 25 December.

People's Daily (2008b) Wang Yang: Zhusanjiao yao shixian tenglong huanniao [PRD must attain 'empty the cage and change the birds'], 17 October.

People's Daily (2009) Guangdong's foreign trade and FDI improve in H1, 3 August.

People's Daily (2010) Chongqing promoted to national central city. 10 February. Accessed on April 12, 2012 at: http://english.peopledaily.com.cn/90001/90776/90882/6892862.html.

People's Daily (2012) Zou Guangdong tese de kexuefazhan zhilu. [To take a scientific developmental path with Guangdong's special characteristics]. 24 August.

People's Daily (2013) Haobu dongyao jianchi he fazhan Zhongguo tese shehuizhuyi, zai shixianzhong buduan yousuo faxian yousuo chuangzao yousuo qianjin. [The unhesitating insistence and development of socialism with Chinese characteristics, to discover, innovate and advance through practice]. 6 January.

People's Daily (2015) Gushi weiji ganyu shi guoji guanli [Intervention in stock market crisis is an international convention]. 20 July.

Perkins, D. and Yusuf, S. (1984) *Rural Development in China*. Baltimore: Johns Hopkins University Press.

Peters, B. G., Pierre, J. and King, D. S. (2005) The politics of path dependency: political conflict in historical institutionalism. *The Journal of Politics*, 67, 1275–1300.

Poncet, S. (2005) A fragmented China: measure and determinants of Chinese domestic market disintegration. *Review of International Economics* 13(3), 409–430.

Prybyla, J. (1967) Red China in motion: A non-Marxist view. In Harry G. Shaffer (ed.) *The Communist World: Marxist and Non-Marxist Views, Volume 2*. New York: Meredith Publishing, pp. 151–183.

Qian, Y. and Roland, G. (1998) Federalism and the soft budget constraint. *American Economic Review*, 88(5), 1143–1162.

Qian, Y. and Weingast, B. (1995) China's transitions to markets: market preserving federalism, Chinese style.. *Essays in Public Policy: No. 55*. Stanford: Hoover Institution Press.

Qin, H. (2004) *Shixian Zhiyou [Freedom of praxis]*. Hangzhou: Zhejiang Renmin Chubanshe.

Qin, H. and Jin, Y. (2012) *Shinian cangsang: Dongou zhuguo de jingjishehui zhuangui yu sixiang bianqian [10 Years of Anguish: Socioeconomic and Philosophical Changes in Various Eastern European Countries]*. Beijing: Dongfang Chubanshe.

Qiu, F. (2011) "Qiu Feng: Guangdong moshi yu Chongqing moshi de zhengzhi yihan" [Qiu Feng: The political implications of the Guangdong and Chongqing models]. Accessed 15 April 2013 at: http://www.aisixiang.com/data/detail.php?id=46068.

Qu, H. (2012) "Qu Hongbin: Guangdong moshi yu Chongqing moshi bijiao" [Qu Hoingbin: comparison of the Guangdong and Chongqing 'models']. Accessed on 15 April 2013 at: http://comments.caijing.com.cn/2012-05-04/111837075.html.

Rawski, T. G. (1995) Implications of China's reform experience. *The China Quarterly*, 144, 1150–1173.

Remick, E. J. (2004) *Building Local States: China During the Republican and Post-Mao Eras*, Vol. 233. Cambridge: Harvard University Press.

Reuters (2012) Hong Kong to loosen yuan conversion limits in further FX reform. 25 July.

Reuters (2013) Red tape hinders Qianhai cross-border yuan loan scheme. 15 August.

Reuters (2015) Beijing's stock rescue has $800 billion bark, small market bite. 23 July.

Rhodes, R. A. W. (1997) *Understanding Governance*. Buckingham: Open University Press.

Rodrik, D. (1998) Who needs capital account convertibility?. *Princeton Essays in International Finance*, 207, 55–65.

Rozelle, S., Park, A., Huang, J. and Jin H. (2000) Bureaucrat to entrepreneur: the changing role of the state in China's grain economy. *Economic Development and Cultural Change*, 48(2), 227–252.

Sanderson, H. and Forsythe, M. (2013) *China's Superbank: Debt, Oil and Influence: How China Development Bank is Rewriting the Rules of Finance*. Singapore: Wiley.

Sang, B. X. (1993) Pudong: Another Special Economic Zone in China – An Analysis of the Special Regulations and Policy for Shanghai's Pudong New Area. *Northwestern Journal of International Law and Business*, 14, 130–160.

Schenk, C. (2002) *Hong Kong as an International Financial Centre: Emergence and Development, 1945–1965*. London: Routledge.

Schenk, C. R. (2007) Economic and Financial Integration between Hong Kong and Mainland China before the Open Door Policy 1965–75. Working paper available at SSRN 1026322.

Segal, A. and Thun, E. (2001) Thinking globally, acting locally: local governments, industrial sectors, and development in China. *Politics & Society*, 29(4), 564–588.

Shambaugh, D. (Ed.). (2000) *The Modern Chinese State*. Cambridge: Cambridge University Press.

Shen, J. (2007) Scale, state and the city: urban transformation in post-reform China. *Habitat International*, 31, 303–316.

Shen, C., Jin, J. and Zou, H. F. (2012) Fiscal decentralization in China: history, impact, challenges and next steps. *Annals of Economics & Finance*, 13(1), 1–51.

Sheng, Y. (2010) *Economic Openness and Territorial Politics in China*. New York: Cambridge University Press.

Shevchenko, A. (2004) Bringing the party back in: the CCP and the trajectory of market transition in China. *Communist and Post-Communist Studies*, 37(2), 161–185.

Shih, V. C. (2008) *Factions and Finance in China*. Cambridge: Cambridge University Press.

Shiraev, E. and Yang, Z. (2014) The GAO-RAO affair: a case of character assassination in Chinese politics in the 1950s. In: *Character Assassination Throughout the Ages*, M. Icks and E. Shiraev (Eds). pp. 237–252). Palgrave Macmillan US.

Shirk, S. L. (1993) *The Political Logic of Economic Reform in China*. Berkeley: University of California Press.

Smart, A. and Lin, G. (2007) Local capitalisms, local citizenship and translocality: rescaling from below in the Pearl River Delta Region, China. *International Journal of Urban and Regional Research*, 31, 280–302.

Snead, W. G. (1975). Self-reliance, internal trade and China's economic structure. *The China Quarterly*, 62, 302–308.

So, A. Y. (2003) The changing pattern of classes and class conflict in China. *Journal of Contemporary Asia*, 33(3), 363–376.

Solinger, D. (1989) Capitalist measures with Chinese characteristics. *Problems of Communism*, 38(1), 19–33.

Solinger, D. (1999) China's floating population: implications for state and society. In: R. MacFarquhar and M. Goldman (eds.), *The Paradox of China's post-Mao Reforms*, pp. 220–240. Cambridge, MA: Harvard University Press.

Su, F. and Yang, D. L. (2000) Political institutions, provincial interests, and resource allocation in reformist China. *Journal of Contemporary China*, 24(9), 215–230.

Sun, M. and Fan, C. C. (2008) Regional Inequality in the Pan-Pearl River Delta Region. In *The Pan-Pearl River Delta: An Emerging Regional Economy in Globalizing China*. Edited by Y.M. Yeung and J. Shen. Hong Kong: The Chinese University Press, pp. 241–269.

South China Morning Post (2012) Chongqing model 'barrier to reforms'. March 26.

South China Morning Post (2013) Hengqin begins fast-paced ride to development. 2 September.

South China Morning Post (2017a) Chongqing still grappling with 'pernicious legacy' of Bo Xilai. 14 February.

South China Morning Post (2017b) How the Communist Party controls China's state-owned industrial titans. 17 June.

Strauss, J. (1998). *Strong Institutions in Weak Polities: State Building in Republican China, 1927–1940*. Oxford: Oxford University Press.

Strauss, J. (2006) Morality, coercion and state building by campaign in the early PRC: regime consolidation and after, 1949–1956. *The China Quarterly*, 188, 891–912.

Su, X. (2012a) Rescaling the Chinese state and regionalization in the Great Mekong Subregion. *Review of International Political Economy*, 19, 501–527.

Su, X. (2012b) Transnational regionalization and the rescaling of the Chinese state. *Environment and Planning A*, 44, 1327–1347.

Sun, Y. and Chan, R. C. (2017) Planning discourses, local state commitment, and the making of a new state space (NSS) for China: evidence from regional strategic development plans in the Pearl River Delta. *Urban Studies*, 54(14), 3281–3298.

Swyngedouw E. (1997) Neither Global Nor Local: 'Glocalisation' and the Politics of Scale. In Cox, K. (Ed.) *Spaces of Globalization: Reasserting the Power of the Local*. New York/London: Guilford/Longman, pp. 137–166.

Teets, J. C. and Hurst, W. (2014) *Local Governance Innovation in China: Experimentation, Diffusion, and Defiance*. Abingdon: Routledge.

Teiwes, F. C. (1990) *Politics at Mao's Court: Gao Gang and Party Factionalism in the Early 1950s*. Armonk: M.E. Sharpe.

Teiwes, F. C. (1993) *Politics and Purges in China: Rectification and the Decline of Party Norms, 1950–1965*. 2nd Edition. Armonk: ME Sharpe.

Tencent Finance (2015) Guangdong zimaoqu luodi beihou: cong bizhao Shanghai dao jinrong jingzheng [Behind the actualization of Guangdong Free Trade Zone: from contrasting Shanghai to financial competition]. 10 May. Available online at: http://finance.qq.com/a/20150510/020166.htm.

The Diplomat (2013, May 14). China's urban dream denied. Accessed on 26 September 2013 at https://thediplomat.com/2013/05/chinas-urban-dream-denied/?allpages=yes.

The Guardian (2015) China lowers growth target to 7% as it fights 'deep-seated' economic problems. 8 March.

Tomlinson, J. (2012) From 'distribution of industry' to 'local Keynesianism': the growth of public sector employment in Britain. *British Politics*, 7(3), 204–223.

Tsai, K. S. (2004) Off balance: the unintended consequences of fiscal federalism in China. *Journal of Chinese Political Science*, 9(2), 1–26.

Tsing, Y. (2010) *The Great Urban Transformation: Politics of Land and Property in China*. Oxford: Oxford University Press.

Tsui, K. Y. (1991) China's regional inequality, 1952–1985. *Journal of Comparative Economics*, 15(1), 1–21.

Tubilewicz, C. and Jayasuriya, K. (2015) Internationalisation of the Chinese subnational state and capital: the case of Yunnan and the greater Mekong subregion. *Australian Journal of International Affairs*, 69, 185–204.

Villaverde, J., Maza, A., and Ramasamy, B. (2010) Provincial disparities in post-reform China. *China & World Economy*, 18(2), 73–95.

Whiting, S. (2001) *Power and Wealth in Rural China: The Political Economy of Institutional Change*. Cambridge: Cambridge University Press.

Xi, J. 2013 Speech at the Third Plenary Session of the Eighteenth Central Committee of the CPC. Accessed: http://news.qq.com/a/20130102/000088.htm.

Xinhua (2016) Xi Jinping: Jujiao fali guanche wuzhongquanhui jingshen, quebao ruqi quanmian jiancheng xiaokang shehui [Xi Jinping: Focus on implementing the spirit of the fifth plenary session to ensure the timely construction of a well-off society]. 18 January. Available online at: http://news.xinhuanet.com/politics/2016-01/18/c_1117813533.htm.

Walder, A. G. (1995) China's transitional economy: interpreting its significance. *The China Quarterly*, 144, 963–79.

Walder, A.G. (2015) *China under Mao: A Revolution Derailed*. Cambridge, MA: Harvard University Press.

Wallace, J. (2014). *Cities and Stability: Urbanization, Redistribution, and Regime Survival in China*. Oxford: Oxford University Press.

Walter, C. E. and Howie, F. J. (2011) *Red Capitalism: The Fragile Financial Foundation of China's Extraordinary Rise*. Wiley, Singapore.

Wang, S. (2002). Social equity is also the absolute principle. Global Media Journal. November 21. Accessed on April 30, 2012 at: http://www.cuhk.edu.hk/gpa/wang_files/EquityAbsolute.pdf.

Wang, X. (2010) Woguo Shouru Fenpei Xianzhuang, Qushi ji Gaige Sikao [The current state of our country's income distribution, trends and reform reflections]. *Zhongguo Shichang*, 20, 8–19.

Wang, L. and Shen, J. (2016a). Spatial planning and its implementation in provincial China: a case study of the Jiangsu region along the Yangtze River plan. *Journal of Contemporary China*, 25, 669–685.

Wang, L. and Shen, J. (2016b) Spatial planning and its implementation in provincial China: a case study of the Jiangsu region along the Yangtze River plan. *Journal of Contemporary China*, 25, 669–685.

Wang, S. and Hu, A. (1999) *The Political Economy of Uneven Development: The Case of China*. Armonk: ME Sharpe.

Wang, L., Wong, C., & Duan, X. (2016). Urban growth and spatial restructuring patterns: The case of Yangtze River Delta Region, China. *Environment and Planning B: Planning and Design*, 43(3), 515–539.

Ward, K. and Jonas, A. E. G. (2004). Competitive city-regionalism as a politics of space: a critical reinterpretation of the new regionalism. *Environment and Planning A*, 36, 2119–2139.

Webber, M. J. (2012) *Making Capitalism in Rural China*. Cheltenham: Edward Elgar.

Wedeman, A. (2001) Incompetence, noise, and fear in central-local relations in China. *Studies in Comparative International Development*, 35(4), 59–83.

Wedeman, A. (2003) *From Mao to Market: Rent Seeking, Local Protectionism, and Marketization in China.* Cambridge: Cambridge University Press.

Wei, Y. (1996). Fiscal systems and uneven regional development in China, 1978–1991. *Geoforum,* 27(3), 329–344.

Wei, Y. D. (2007). Regional development in China: transitional institutions, embedded globalization, and hybrid economies. *Eurasian Geography and Economics,* 48(1), 16–36.

Wei, C. X. G. (2011) Mao's legacy revisited: its lasting impact on China and post-Mao era reform. *Asian Politics & Policy,* 3(1), 3–27.

Wei, Y. D. (2013) *Regional Development in China: States, Globalization and Inequality.* London: Routledge.

Wen, T. (2011). Chongqing jingyan yi er san [Chongqing experience one two three] Renmin luntan, November, 65–66.

Wenweipo (2009) Wang Yang gaikou: Tenglong huanniao zhishi daoxiang, bu qiangzhi qiye zhuanyi [Wang Yang's change of words: emptying the cage is only a guidance, firms not compelled to relocate]. 12 February.

Wenweipo (2013) Wunian tenglong huanniao, yue 8 wan qiye guanting [5 years of emptying the cage and changing the birds, 80000 enterprises shut down]. 26 January.

Whyte, M. K. (1996) City versus countryside in China's development. *Problems of Post-Communism,* 43(1), 9–22.

Whyte, M. K. (2010) The paradoxes of rural-urban inequality in contemporary China. In: M. K. Whyte (ed.) *One Country, Two Societies: Rural–Urban Inequality in Contemporary China,* pp. 1–25. Cambridge: Harvard University Press.

World Bank (1995) *Macroeconomic Stability in a Decentralized Economy.* Washington, DC: The World Bank.

Wu, F. (2009) Land development, inequality and urban villages in China. *International Journal of Urban and Regional Research,* 33(4), 885–889.

Wu F (2015) *Planning for Growth: Urban and Regional Planning in China.* New York: Routledge.

Wu, F. (2016) China's emergent city-region governance: a new form of state spatial selectivity through state-orchestrated rescaling. *International Journal of Urban and Regional Research,* 40, 1134–1151.

Wu, J. and Ma, G. (2012) Zhongguo Jingji Gaige Ershi Jiang [20 Conversations on China's Socioeconomic Reforms] Beijing: Sanlian Shudian.

Wu, F. and Zhang, J. (2007) Planning the competitive city-region the emergence of strategic development plan in China. *Urban Affairs Review,* 42(5), 714–740.

www.ce.cn (2015) Zhongguo chengxiang jumin shourubi 13 nian lai shouqi shuoxiao zhi 3 bei yixia [China's Urban-rural income disparity ratio goes under 3 for the first time in 13 years]. 20 January. Accessed on 13 March 2015 at: http://www.ce.cn/xwzx/gnsz/gdxw/201501/20/t20150120_4384230.shtml.

www.china.com.cn (2009) "Dubuzhu de nongmingong jincheng hongliu" [The unstoppable urban entry of peasant workers], 5 February. Accessed on 10 September 2013 at: http://www.china.com.cn/news/txt/2009-02/05/content_17230385_3.htm.

www.chinanews.com (2011) "Hu Jintao: fazhan shi ying daoli, wending shi ying renwu", 1 July. Accessed on 30 April 2012 at: http://www.chinanews.com/gn/2011/07-01/3150785.shtml.

www.chinanews.com (2013) "Dazhao 'tequzhong de tequ', Shenzhen wei Qianhai zhaoshang 'xuan' zi", 8 September. Accessed on 27 November 2013 at: http://www.chinanews.com/df/2013/09-08/5259109.shtml.

www.cqnews.net (2012a) "Chongqing caizhengjü renshi cheng Chongqing bucunzai caizheng chizi", 24 March.

www.cqnews.net (2012b) "Chongqing faxing 50yi gongzufang jianshe qiye zhaiquan quanguo guimo zuida", 26 April.

www.dwnews.com (2016) Xi Jinping B20 fenhui yanjiang (quanwen). [Full text of Xi Jinping's speech at the B20 Summit]. 3 September. Available at: http://china.dwnews.com/news/2016-09-03/59766181.html.

www.eastday.com (2011) "2010 nian Shanghai chengshi jumin jiating renjun kezhipei shouru da 31838 yuan. 25 January. Retrieved on 5 May 2012 from: http://sh.eastday.com/qtmt/20110125/u1a850835.html.

www.people.com.cn (2000) Deng Xiaoping: Jianchi sixiang jiben yuanze [Deng Xiaoping: Emphasise the four cardinal principles]. 29 December. Available at: http://www.people.com.cn/GB/channel1/10/20000529/80791.html.

www.news.cn (2012) Huang Qifan jieshou Xinhuawang zhuanfang tan 'Chongqing tansuo [Huang Qifan talks about Chongqing explorations during interview with Xinhuawang] 8 March. Retrieved on 12 April 2012 from: http://news.sina.com.cn/c/2012-03-08/1752 24082541_3.shtml.

Xia, Y. and Feng, Z. (1982) Tidu Lilun yu Quyu Jingji [Ladder-step theory and regional economies]. *Shanghai Kexue Yanjiusuo Qikan (Yanjiu yu Jianyi)*, 8, 21–24.

Xinhua (2009) Zhongguo jishi: 1958 nian chengxiang eryuan huji zhidu queli [Remembering China: Urban rural dual structure hukou institution established in 1958]. 13 August.

Xinhua (2012) "Yanghang fuhangzhang: Quyuxing jinrong gaige qieji yihongershang", 24 November.

Xinhua (2013a) "Reports on "China's rejection of urbanization plan" untrue: senior official", 24 May.

Xinhua (2013b) Huanquan fabu: Zhongguo gongchandang dishibajie zhongyang weiyuanhui disanci quanti huiyi gongbao [Authorized release: Communique of the Third Plenum of the CPC Central Committee], 12 November.

Xinhua (2014a) "2013 nian zhongguo chengxiang shourubi 3.03: 1 wei 10 nian lai zuidi" [2013 urban-rural income ratio 3.03, lowest in 10 years], 20 January.

Xinhua (2014b) Chongqing Liangjiang Xinqu 2013 nian waimao jinchukou zhongzhi da 305.1 yi meiyuan [Import-export external trade for Liangjiang New Area reached US$30.5 billion in 2013] 22 January.

Xinhua (2014c) Huanquan fabu: Zhongguo gongchandang dishibajie zhongyang weiyuanhui disanci quanti huiyi gongbao [Authorized release: Communique of the Third Plenum of the CPC Central Committee], 12 November.

Xinxi Shibao (2009a) Wang Yang jingcai yulu. [Speech highlights of Wang Yang]. 18 July.

Xinxi Shibao (2009b) Dongguan chanye jiegou tiaozheng, bingfei ganchang ganren. [Economic restructuring in Dongguan is not about chasing away firms and people]. 29 July.

Xu, C. (2011) The fundamental institutions of China's reforms and development. *Journal of Economic Literature*, 49, 1076–1151.

Xu, Y. (2013) Part of academic discussion at the Unirule Institute of Economics, Beijing, 6 December 2012; transcript published in China Review (14 January 2013); author's translation.

Yang, C. (2005). Multilevel governance in the cross-boundary region of Hong Kong–Pearl River Delta, China. *Environment and Planning A*, 37(12), 2147–2168.

Yang, J. (2012) *Tombstone: the Great Chinese Famine*. London: Penguin.

Yang, R. and He, C. (2014) The productivity puzzle of Chinese exporters: perspectives of local protection and spillover effects. *Papers in Regional Science*, 93(2), 367–384.

Yang, C. and Li, S. M. (2013) Transformation of cross-boundary governance in the Greater Pearl River Delta, China: contested geopolitics and emerging conflicts. *Habitat International*, 40, 25–34.

Yangcheng Wanbao (2009) Hu Jintao shangwu dao Guangdongtuan shenyi baogao, qinting daibiao yijian. [Hu Jintao considers the Guangdong delegation's report in the morning, listens to representatives' opinions], 7 March.

Yangcheng Wanbao (2013) Nansha 15 xiang jinrong zhengce huopi, fazhan qianli chao Qianhai Hengqin [15 financial policies approved in Nansha, developmental potential exceeds Qianhai & Hengqin], 13 November.

Yep, R. (2008) Enhancing the redistributive capacity of the Chinese state? Impact of fiscal reforms on county finance. *The Pacific Review*, 21(2), 231–255.

Yeung, Y. M., Lee, J. and Kee, G. (2009) China's special economic zones at 30. *Eurasian Geography and Economics*, 50(2), 222–240.

Young, A. (2000) The Razor's edge: distortions and incremental reform in the people's republic of China. *Quarterly Journal of Economics*, 115(4), 1091–1135.

Yue, E. (2012) "Renminbi: A New Era and the Role of Hong Kong". Keynote Address at Euromoney Global Offshore RMB Funding Forum 2012. Accessed on 10 January 2014 at:http://www.hkma.gov.hk/eng/key-information/speech-speakers/ewmyue/20120523-1.shtml.

Zeng, D. Z. (2010) Building Engines for Growth and Competitiveness in China: Experience with special economic zones and industrial clusters. ISBN: 978-0-8213-8432-9. e-ISBN: 978-0-8213-8433-6 https://doi.org/10.1596/978-0-8213-8432-9.

Zeng, J. (2015) Did policy experimentation in china always seek efficiency? A case study of Wenzhou financial reform in 2012. *Journal of Contemporary China*, 24(92), 338–356. doi: https://doi.org/10.1080/10670564.2014.932517

Zeng, J. (2016) *The Chinese Communist Party's Capacity to Rule: Ideology, Legitimacy and Party Cohesion*. Basingstoke: Palgrave Macmillan.

Zhang, J. (2013) Marketization beyond neoliberalization: a neo-Polanyian perspective on China's transition to a market economy. *Environment and Planning A*, 45(7), 1605–1624.

Zhang, J. and Peck, J. (2016). Variegated capitalism, Chinese style: regional models, multi-scalar constructions. *Regional Studies*, 50(1), 52–78.

Zhang, X. Q. (1997). Chinese housing policy 1949–1978: the development of a welfare system. *Planning Perspectives*, 12(4), 433–455.

Zhang, Y., Yan, F. and Wang, Z. (2015) Kangzhan shiqi binggong neiqian Chongqing de jingji yingxiang [On the economic impact of the inward migration of military industries during the anti-war period]. *Journal of Chongqing Technology and Business University*, 32(2), 106–111.

Zhao, Z. (2009) *Prisoner of the State: The Secret Journal of Premier Zhao Ziyang*. New York: Simon and Schuster.

Zhongguo Qingnianbao (2004) Difang baohuzhuyi yicheng quanguo tongyishichang de zuida zhangai [Local protectionism has become the biggest barrier to national economic unification]. 23 June.

Zhou, Q. (1995) Zhongguo nongcun gaige: Guojia he suoyouquan guanxi de bianhua (shang) [Reform in China's countryside: Changes to the relationship between the state and ownership (part 1)]. *Guanli Shijie*, 3, 178–220.

Zhou, X. (2012) Zhou Xiaochuan: Woguo jinrong gaige zhong zhixiaershang de zhucheng bufen [Zhou Xiaochuan: The bottom-up constituent parts our country's financial reforms]. Accseed on 15 January 2014 at: http://www.pbc.gov.cn/hanglingdao/128697/128719/128766/2864663/index.html.

Zhou, M. and Logan, J. R. (2002). Market transition and the commodification of housing in urban China. In J.R. Logan (Ed.), *The new Chinese city: Globalization and Market Reform*, pp. 135–152. Oxford: Blackwell.

Zhu, Z. (2007) Reform without a theory: Why does it work in China? *Organization Studies*, 28(10), 1503–1522.

Zhu, R. (2011) *Zhu Rongji jianghua shilu [Transcripts of Zhu Rongji's Speeches]*. Beijing: Renmin Chubanshe.

Zhu, J. (2013) *Zhonguo gaige de qilu [The Crossroads of Reforms in China]*. Taipei: Linkingbooks.

Zhuhai Tequbao (2016) Hengqin Xinqu leiji wancheng gudingzichan touzi 1147yi yuan [Hengqin New Area accumulated fixed capital investments of 114.7 billion *yuan*], 11 March.

Zhuhai Statistics Information Network. Accessed at: http://www.stats-zh.gov.cn.

Index

NOTE: Page numbers in *italic* type refer to figures, page numbers in **bold** type refer to tables.

On Shifting Foundations: State Rescaling, Policy Experimentation and Economic Restructuring in Post-1949 China, First Edition. Kean Fan Lim.
© 2019 Royal Geographical Society (with the Institute of British Geographers).
Published 2019 by John Wiley & Sons Ltd.